WATER STRESS
ON PLANTS

WATER STRESS
ON PLANTS

G. M. Simpson

PRAEGER

PRAEGER SPECIAL STUDIES • PRAEGER SCIENTIFIC

Library of Congress Cataloging in Publication Data

Simpson, G. M.
    Water stress on plants.

    Bibliography: p.
    Includes index.
    Contents: Global perspective on drought / G.M. Simp-
son—The development of water deficits in plants /
A. J. Karamanos—The responses of plants to drought
stress / J. M. Clarke and R. C. Durley—[etc.]
    1. Plants, Effect of drought on.   2. Plant-water
relationships.   3. Plants, Effect of stress on.   I. Ti-
tle.
    QK754.7.D75S55   581.2′4   81-7376
    ISBN 0-03-056698-3   AACR2

33,848

Published in 1981 by Praeger Publishers
CBS Educational and Professional Publishing
A Division of CBS, Inc.
521 Fifth Avenue, New York, New York 10175 U.S.A.

© 1981 by Praeger Publishers

123456789   145   987654321

Printed in the United States of America

PREFACE

The study of the effects of drought stress on plant water relations, like most other fields of biological research, has expanded exponentially in the last thirty years. The number of journal articles, books, and reports produced in this field in one year now approximates the total number produced over the first half of this century. Notwithstanding this volume of research, the effects of water stress on plants continue to have a prominent place in the day-to-day experience of farmers and nomadic people all around the globe. All segments of society suffer hardship when periodic severe droughts bring economic and social disaster to different regions of the globe.

Since ancient times irrigation has been the most important practical means of offsetting regular drought periods. However, in this century the growth of knowledge in the fields of soil science, plant physiology, and plant genetics has opened up new possibilities for the ways in which plants can be grown under a range of drought conditions.

The objective of this book is to present a synopsis of the current extent of research into the effects of water stress on plants. It is directed particularly to the general reader, as well as to students of biology and agriculture. Because much of the basic research on plant water relations is sophisticated biophysics utilizing a highly specialized vocabulary and complicated techniques, the emphasis in this book has been placed on using a simplified and general approach rather than an intensive treatment of the subject matter. In this way it is hoped that a holistic view of drought-stress research in plants is projected that can be easily related to agriculture, plant breeding, and the food, fiber, energy, and demographic problems facing mankind.

CONTENTS

CONTRIBUTORS

John M. Clarke

Experiment Station
Agriculture Canada
Swift Current, Saskatchewan
Canada

Richard C. Durley

Crop Science Department
University of Saskatchewan
Saskatoon, Saskatchewan
Canada

Andreas J. Karamanos

Laboratory of Plant Physiology
The Agricultural College of Athens
Votanikos, Athens,
Greece

Graham M. Simpson

Crop Science Department
University of Saskatchewan
Saskatoon, Saskatchewan
Canada

ACKNOWLEDGMENTS

Thanks are due to our patient families, and to our respective institutions, for cooperation during the preparation of the manuscript.

Special thanks are due to John Diduck for graphic work, to Mrs. Ann Diament for typing the manuscript, and to Bob Redmann for advice. The authors also thank the scientists who gave permission to use figures from previously published work.

1

Global Perspectives
on Drought

GEOGRAPHICAL ZONES

In a book concerned primarily with plants, the geographical perspective on drought could be defined in terms of water deficits related to plant growth. However, the plant viewpoint is too narrow for gaining an overall view of the dimensions of drought since drought affects agriculture generally and also the social well-being and future of humanity. Drought has different meanings according to the perspective adopted. Meteorologists see it as a negative departure from normal precipitation, geographers see the need for mapping global climatic descriptions, hydrologists see areas of underground water depletion, and agronomists measure drought as a loss of crop and animal potential with marked social consequences. Thus, in looking at the geographical distribution of drought to obtain a global perspective, some measure of each of these viewpoints is necessary for an appreciation of the full dimensions of drought.

Using average annual precipitation modified by its form of distribution during the year combined with a temperature coefficient, the driest areas of the globe can be classified into zones of aridity that demarcate the long-term climatic conditions. Meigs' classification

(Meigs 1953) is such a description (Figs. 1, 2). These divisions will also broadly define the kinds of natural vegetation that are likely to exist within the boundaries. At any point in time the actual vegetation may or may not represent the climax vegetation that would occur in the absence of major interventions such as wide-scale volcanic activity or the activities of man. There are several classification systems relating natural vegetation to the climatic conditions. Examples are the empirical approach of Köppen (1931) and the methods of Raunkaier (1934), Thornthwaite (1939, 1948), and Penman (1949), which use meteorological parameters to define plant zones without reference to the topography or soils. The nature of soils is fairly well correlated with long-term climatic conditions.

These climatic classifications, based on long-term annual data, can broadly define the expected annual conditions and thus the general groupings of plants adapted to these zones. However, the aspect of aridity that has the greatest effect on the long-term sustainability of plant productivity, on which the future of man depends, is the long- and short-term perturbations in rainfall that occur with a randomness that frequently brings calamity in its wake for large sections of mankind.

Long-term climatic changes have been important in determining the activity and longevity of various civilizations. Since neolithic times these climatic changes have interacted with the activities of both nomadic and agricultural man to create an accelerated process of desertification (Secretariat U.N. 1977). These periods of changed climate have often been long enough to cause the demise of cultures that not only could not adapt but actually accelerated the natural change through overgrazing of natural grasslands, wanton destruction of forests, and prolonged arable agriculture in areas susceptible to wind and water erosion and soil salinization (Nir 1974; Secretariat U.N. 1977).

The frequency and persistence, often for tens of thousands of years, of these long-term periods of aridity is illustrated in Fig. 3 for the nine major regions of the globe that presently have major deserts or extensive arid to semiarid areas.

Rumney's (1968) classification of the deviation from annual average of world precipitation (Fig. 4) indicates that in general the zones with the lowest rainfall tend to have the highest deviation from normal. Nevertheless within any geographical zone, even a zone of high rainfall, there is the possibility of and a randomness to the occurrence of drought that can have widespread and disastrous effects on plant life. This in turn limits all forms of food production.

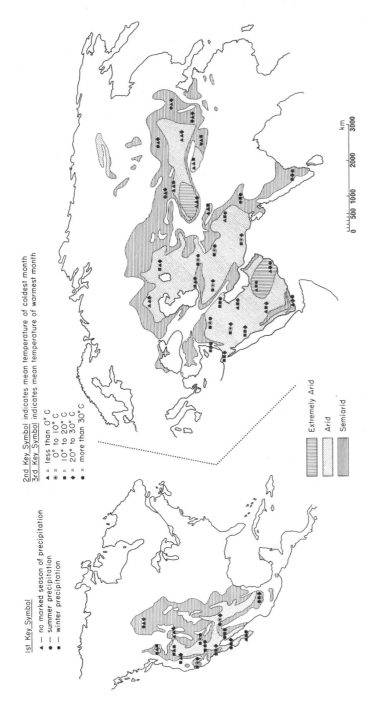

1st Key Symbol
▲ — no marked season of precipitation
● — summer precipitation
■ — winter precipitation

2nd Key Symbol indicates mean temperature of coldest month
3rd Key Symbol indicates mean temperature of warmest month

▲ = less than 0° C
◉ = 0° to 10° C
■ = 10° to 20° C
♦ = 20° to 30° C
● = more than 30° C

Extremely Arid

Arid

Semiarid

0 500 1000   2000   3000
km

Fig. 1. Arid lands of North America and Asia (after Meigs).

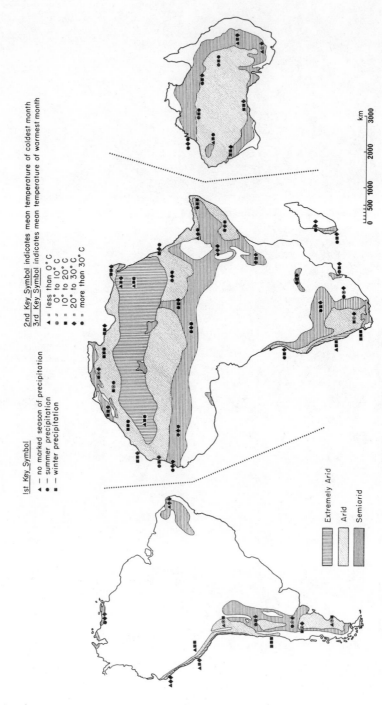

1st Key Symbol

▲ – no marked season of precipitation
● – summer precipitation
■ – winter precipitation

2nd Key Symbol indicates mean temperature of coldest month
3rd Key Symbol indicates mean temperature of warmest month

▲ = less than 0° C
⊛ = 0° to 10° C
■ = 10° to 20° C
◆ = 20° to 30° C
● = more than 30° C

Extremely Arid

Arid

Semiarid

km

0   500 1000   2000   3000

Fig. 2. Arid lands of South America, Africa, and Australia (after Meigs).

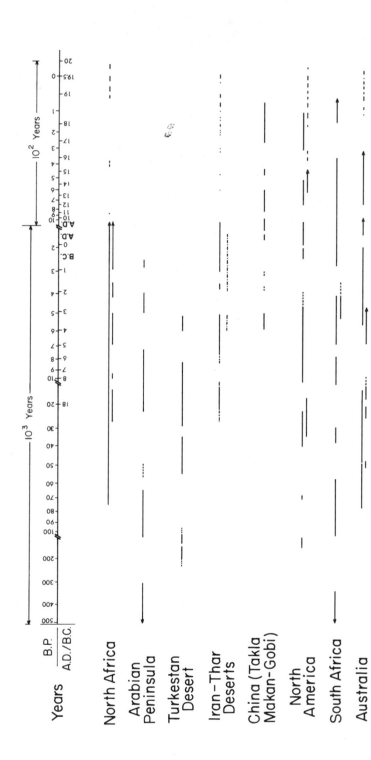

Fig. 3.   Periods of aridity that have influenced the major land regions subject to drought (after Secretariat U.N. 1977).

/   5

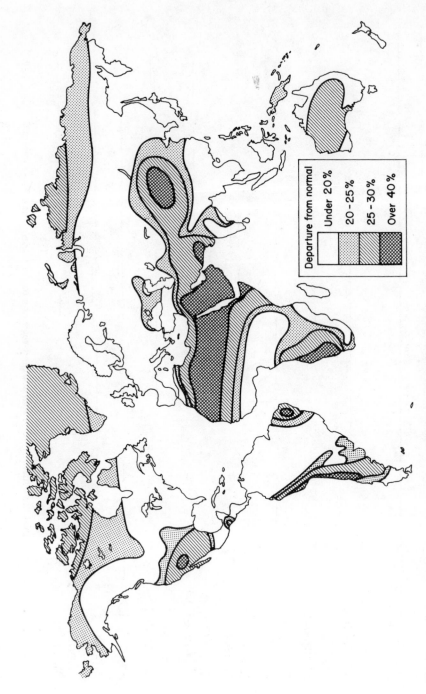

Fig. 4. Deviation from annual average of world precipitation (after Rumney 1968).

Departure from normal

Under 20%
20-25%
25-30%
Over 40%

TABLE 1: The Number of Droughts Per Century in China during Different Dynasties

Dynasty	Tang	Fifth Dynasty and North Sung	South Sung	Yuen	Ming	Manchu
Christian Era	618–907	908–1126	1127–1279	1280–1367	1368–1643	1644-1847 1861-1900
Capital	Chang-an Shensi	Kai-fung Honan	Hangchow Chekiang	Peking Chihli	Peking Chihli	Peking Chihli
Honan	4.2	17.8	1.3	34.4	2.2	26.0
Chihli	2.1	6.9	3.9	25.3	1.8	43.7
Shensi	9.1	1.8	3.9	4.6	2.2	11.6
Shansi	0.7	2.3	—	4.6	7.3	12.3
Shantung	1.7	5.5	0.7	20.7	2.2	27.7
Kansu	0.3	1.8	1.3	5.7	—	8.3
Chekiang	1.4	1.4	17.8	4.6	4.0	22.7
Kiangsu	1.4	2.7	9.9	3.4	1.5	43.8
Hupeh	0.3	0.9	4.6	4.6	0.7	26.2
Szechwan	0.7	—	2.6	—	1.1	2.9
Anhwei	0.7	3.7	5.9	4.6	—	36.3
Kiangsi	0.7	1.4	5.9	4.6	1.5	21.8
Hunan	—	1.4	—	3.4	1.1	20.6
Fukien	—	0.9	4.6	4.6	3.3	6.5
Kwangsi	—	0.5	—	1.2	0.7	1.6
Yunnan	—	—	—	—	6.9	2.5
Kweichow	—	—	—	—	—	2.5
Kwangtung	—	0.5	0.7	2.3	1.5	7.0

Source: After Mallory (1926).

For example, in the last 1,400 years China has recorded a serious drought on a number of occasions in every major province (Table 1). Similarly, in India, currently the next most highly populated country after China, the likelihood of drought occurring in any province—once in every 5 or 10 years—is relatively high (Table 2).

The short-term fluctuations of rainfall, on a weekly or monthly basis associated with a crop cycle, can determine the success or failure of specific crops, particularly in the arid and semiarid zones. Poor timing of rainfall in relation to different growth stages can cause crop failure even when overall annual rainfall exceeds the

TABLE 2: Frequency of Occurrence of Droughts in India

Areas	Frequency of Occurrence
1. Northeastern regions	Once in 15 years
2. Bihar, West Bengal, Orissa, Coastal Andhra Pradesh, Kerala, Coastal Karnataka and Konkan and East Madhya Pradesh and Maharashtra	Once in 5 years
3. Vidarbha, Uttar Pradesh, North Interior Karnataka, West Madhya Pradesh	Once in 4 years
4. Telengana, Rayalaseema, Tamilnadu, South Interior Karnataka, Eastern Rajasthan and Gujarat	Once in 3 years
5. West Rajasthan and Kutch	Once in 2.5 years

Source: From Das (1972).

normal crop requirement. For example, in the short, one-crop season of approximately 120 frost-free days of the temperate semiarid zone of the northwestern prairie of North America (Saskatchewan, Canada), the rainfall in May, June, and July determines the success or failure of a crop of wheat. The randomness and relatively high frequency of dryness during these periods is illustrated in Fig. 5 for a period of 93 years.

In a more tropical region where several crops per season may be grown, severe water deficits over periods of several weeks can markedly affect crop growth, soil cultivation, and timeliness and success of planting of consecutive or overlapping interrow crops. Again there is randomness coupled with high frequency of these drought periods, which is illustrated in Fig. 6 by records over 67 years at Palamau, India (Singh 1978).

Another important cyclical event, usually quite predictable, is the annual arrival and departure of the monsoons in India and Southeast Asia and the equivalent cyclical rainfall in the sub-Sahara region of Africa. However, occasionally when the monsoons fail to arrive on time, or fail altogether, there can be drought of exceptional severity. For example, in India approximately 70% of the annual rainfall occurs in the 2- to 4-month period of the monsoon (Das 1972). Since planting generally depends on first cultivating the soil after the initial rains have softened the soil, the timeliness of planting depends on the timeliness of the arrival of the monsoon (Fig. 7). On the Indian subcontinent the greatest variability in arrival of the monsoon occurs in the northwest of India and Rajasthan, which are also the areas that receive the smallest amounts of rain (Das 1972). It is generally in these zones of erratic and low rainfall that crop species well adapted to drought by avoidance or tolerance mechanisms are grown.

The consequences of a drought that occurs in an area of well-distributed and adequate rainfall suited for highly productive mesophytic crop plants can of course be very great. However, the frequency of occurrence is relatively low when compared with the semiarid and arid zones.

ECONOMIC PLANT DISTRIBUTION AND HUMAN DISTRIBUTION

Modern investigation of the history of argicultural development, aided by radiocarbon dating and sophisticated techniques of archaeology, have indicated where humanity probably first domesticated plants. There seems little doubt that, in the regions of western Asia and the eastern Mediterranean, cereals were first domesticated in the dry upland areas, beginning about 10,000 years ago (Arnon 1972; Grigg 1974). It is in these zones that the wild progenitors of wheat and barley, as well as those of sheep, goats, cattle, and pigs, were found. In the New World there were two areas in which cereal domestication took place about 6000 B.C.: Meso-America and coastal Peru (Grigg 1974). These areas are similar to the western Asian areas in that there is considerable variation in topography and climate over short distances, which is conducive to the maintenance of ecotype diversity within species. These centers of cultivated plants were first noted by Vavilov (1949–50) in the 1930s. Vavilov's work indicates that the wet tropical areas of Southeast Asia were also important centers of vegeculture. For example, the taro, yam, breadfruit, sago palm, coconut, and banana orginate from there. The earliest archaeological evidence for the legumes, possibly already domesticated, in this area is about 9000 B.C.

Given the circumstances that in tropical regions of high rainfall archaeological evidence and plant remains are likely to be much less persistent than in semiarid or arid regions, it seems possible that the

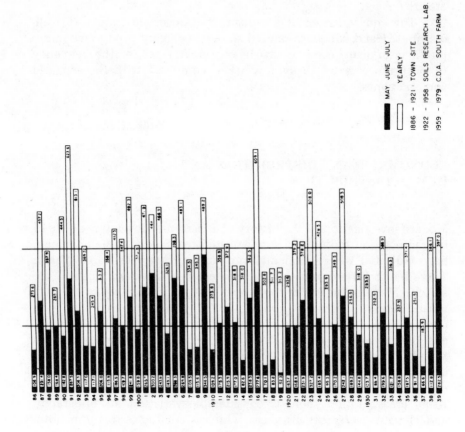

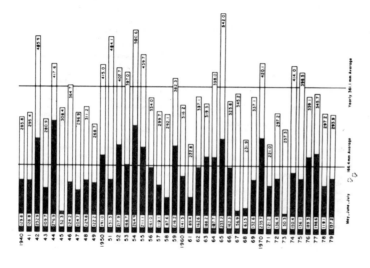

Fig. 5. Precipitation (mm) for the years 1886–1979 at Swift Current, Saskatchewan, Canada.

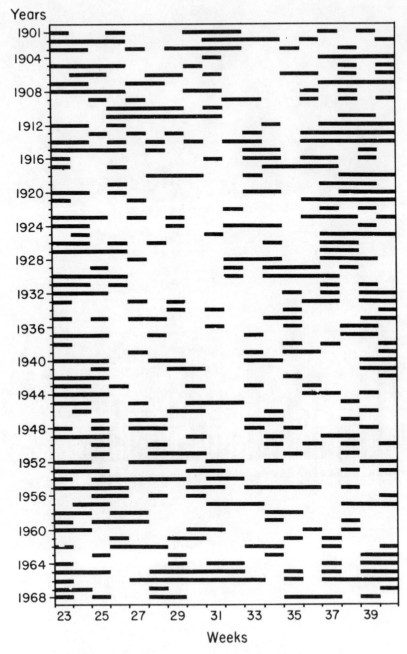

Fig. 6. Distribution of various droughts for Palamau, India (annual, seasonal, and weekly) (after Singh 1978).

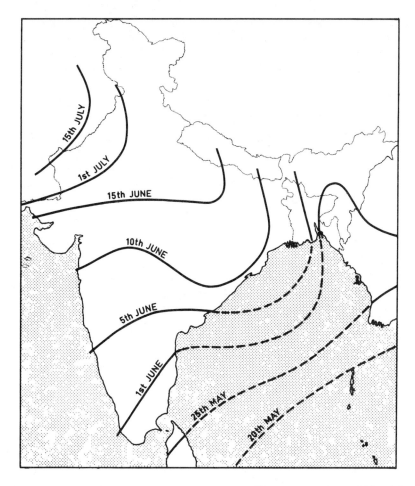

Fig. 7.   Timing of arrival of the monsoon at different latitudes in India (after Das 1972).

argument that agricultural humanity first made the transition from a nomadic existence to arable farming in the semiarid upland areas of western Asia (Arnon 1972) may not be correct. Certainly, vegeculture in Southeast Asia, West Africa, and the uplands of South America preceded seed culture in these areas, and the domestication of animals was unimportant.

It is clear from archaeological evidence that irrigation did not develop until some five or six thousand years after the first domestications of plants in both western Asia and Meso- and South America (Grigg 1974). Stabilizing plant production by the adoption of irrigation technology when combined with the more complex social organization necessary for carrying out and coordinating the varied tasks simultaneously provided the social framework for building the first major civilizations. These civilizations were first found in either the semiarid areas (Tigris, Euphrates) or the arid areas (Nile valley) where plant growth in the absence of irrigation was erratic or possible only for xerophytes.

Most crops grown today were domesticated prior to about 2000 B.C. and were originally confined to a very small part of the land surface. Vavilov (1949–50) estimated that the progenitors (wild types) of modern crops originally occupied less than 10% of the land surface. A comparison of Vavilov's major areas of plant diversity with the current distribution of national and international centers collaborating with the International Board for Plant Genetic Resources (IBPG 1978) indicates that the centers of plant diversity are still of great significance 12,000 years later (Fig. 8).

Until the advent of the mercantile empires of Portugal, Spain, France, Holland, and Great Britain, the diffusion of crops from their centers of origin was relatively slow. By the eighteenth century the major West Asian and Meso-American plants were already scattered around the globe, primarily in areas to which the plants were easily adapted. Most of the cultivated forage grasses originated in the Mediterranean region or northern Eurasia (Grigg 1974) and were transferred to the Americas, South Africa, Australia, and New Zealand by Europeans before the end of the nineteenth century. Thus the modern spread of domesticated plants on a global scale has taken place swiftly over a period of about 500 years, and the process is still highly active (Hawkes et al. 1976) both for the introduction of new species and the genetic improvement of already introduced species. Mangelsdorf estimates that man has used some 3,000 species of plants for food, of which about 150 are of major importance in commerce (Mangelsdorf 1966). Virtually every country in the world now has some formal mechanisms for the introduction of new species or exchange of genetic resources for crop plants. The almost universally grown food crops such as wheat, maize, soybeans, rice, and the plantation crops such as sugarcane, coffee, tobacco, and rubber have been first introduced into geographical and climatic zones roughly similar to the centers of origin. Many of the species that originated in semiarid regions—for example, wheat, barley, millet, and maize—now occupy extensive zones with a more stable and higher rainfall. These new regions were previously occupied by forests or tall grassland associations. Thus removal of moisture as a rate-limiting factor to growth has enhanced yields per hectare closer to the genetic potential of these crops.

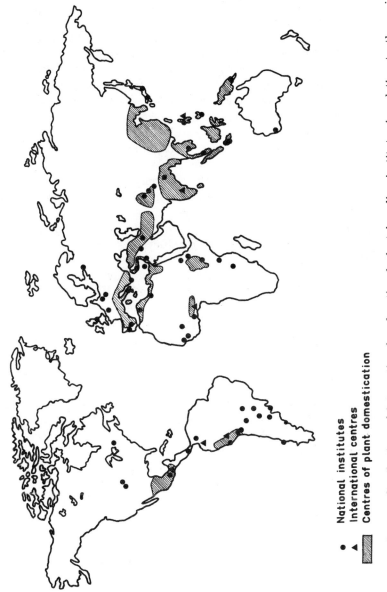

National institutes
International centres
Centres of plant domestication

Fig. 8. Current distribution of international and national plant breeding institutes in relation to the major centers of plant domestication.

/ 15

Human distribution and the rapid expansion of the global population has taken place primarily in the moister lands of the world. For example, only 14% of the world's population live in the drylands and, of this group of about 630 million people, approximately 72% live in the semiarid zone, 27% in the arid zone, and only 1% in the extremely arid zone (Secretariat U.N. 1977). According to Vavilov (1949-50) nearly three-fifths of the total sown area in the world in the 1930s was cultivated with moderately drought-resistant crops such as wheat, barley, corn, rye, flax, sunflower, and sugar beet. Some of these crops have a potentially lower optimum productivity per hectare than the more moisture-dependent crops such as rice, soybeans, oats, beans, peas, and many vegetables and fruits. The adaptation of the moderately drought-resistant crops to a world scale of production is related less to their drought resistance than to other factors such as their annual habit of growth, adaptation to specific growing seasons, function in a crop rotation system, social preference, and export trade.

Humanity has thus achieved high production per unit of land and an overall global increase in crop production by a combination of several practices that achieve a form of drought evasion. First, species adapted to survival under low precipitation have been introduced into areas of higher or more reliable rainfall. This has generally involved replacement of the indigenous vegetation such as forest or grassland. Second, there has been development of various forms of irrigation in semiarid zones which otherwise only support limited crop production under low and/or erratic rainfall. Third, plants have been grown under irrigation in arid zones with the possibility of year-round growth where normal plant growth would be limited to a few slow-growing xerophytes. Last, rain and floodwater conservation have been supplemented in the humid zones with irrigation to increase the level and stability of production of crops such as rice.

It is estimated that currently about 14% (200 million hectares) of the world's arable lands are irrigated and that they use about 1,400 billion cubic meters of water per annum (Barton Worthington 1976). Where irrigation can be practiced, yield increases above those possible with natural precipitation are often spectacular. Tenfold increases in yield are not uncommon, because the full genetic potential of the plant can be realized when water, nutrients, weed competition, and poor planting are eliminated as limiting factors.

The potential significance of irrigation for achieving further increased crop productivity in areas otherwise subject to drought stress can be gauged from the fact that of the world's 3.2 billion hectares of potentially arable land, 2.02 billion hectares are located in the developing regions of Africa, Asia, and America, which also have high human populations. About 1.330 billion hectares of this latter area is potentially irrigable land, and there is water available in these areas sufficient to irrigate 99% of the land if investment resources and manpower could be made available (Obeng 1975).

Wherever irrigation is practiced the increased costs associated with land development, drainage, fertilizers, and pesticide control tend to be covered by growing crops that have a high cash return. Thus a fiber crop such as cotton or maize for livestock may receive preference over food crops for local human consumption. In some areas the long-term side effects associated with faulty irrigation may eventually offset any short-term gains in plant productivity per acre or increased food production on a national scale. In the Euphrates valley, historically one of the oldest sites of irrigation in the world, approximately 50% of the soils are salt-affected and water-logged; in Egypt, the figure is about 30%. In Iraq more than 50% of the lower Rafadain plain suffers from salinity and water-logging, and in Pakistan almost 27 million hectares out of a total of 37 million hectares suffer from salinity, water-logging, or both (Barton Worthington 1976).

India already has a major surface-water development with an installed capacity for irrigating 20 million hectares, with a potential of 65 million hectares in the total geographical area of 328 million hectares. Much of the potentially irrigable area, however, lies within the 90 million hectares seriously affected by wind erosion. Failure to maintain cropping—for example, because of salinization of the soil—could ultimately lead to irreversible loss of arable land through wind and water erosion (Obeng 1975), a fate that much of the Middle East has suffered in the last 6,000 years. California, the most productive vegetable- and fruit-producing state in the United States, is highly dependent on irrigation, and problems associated with salinization of soils are already showing after less than 50 years.

The rapid growth of irrigation as a form of technology closely associated with the spread of new crops into semiarid or arid lands to feed far-distant populations is largely a function of the industrial age, since mechanical pumps are essential to any large-scale irrigation project. It is estimated that in 1800 the global area under irrigation was about 8 million hectares, by the end of the nineteenth century about 40 million hectares, and by 1950 around 100 million hectares. Today the distribution in millions of hectares is: Asia, 14; United States, 26; Soviet Union, 13: Europe, 12; Africa, 10; and Australasia, 1. Globally, the prospect is 300 million hectares by the year 2000 (Secretariat U.N. 1977).

The concentration of the mass of humanity into large metropolitan areas, which has accompanied the industrial revolution— a process that is almost completed in the developed countries and is taking place quickly in the underdeveloped countries—has placed new strains on agriculture in the marginal drought-susceptible zones. As mechanization replaces labor, farm size has increased to reap the benefits of economies of scale. Increased farm productivity in many geographic zones has been obtained by bringing in marginal land, often with disastrous consequences. For example, the application of deep ploughing and "dust mulching" to the semiarid lands of the North American prairies in the 1930s, a practice derived from European experience in a moister climate and with different soils, led to the "dust bowl" and large-scale abandonment of land following massive soil erosion by wind. A similar problem arose in the marginal lands of Kazakhstan in the 1960s. In the northwest temperate prairie of Canada the system of summerfallowing adopted to permit cereal growing in the semiarid, drought-prone climate has led to serious losses of organic matter and nitrogen and an increase in salinity over the last 60 years (Rennie 1978).

Often, marginal lands subject to drought and with low human populations are used for production of crops or animal products that are exported from the region. The vagaries of national and international markets combined with periodic droughts lead to cycles of over- and under-utilization, which can be detrimental to the long-term sustainability of the system. An example of this interaction was to be seen in the recent prolonged drought in the Sahel zone of North Africa. Recent developments of an international trade in beef, through which previously nomadic people were encouraged to settle around water holes and wells adjacent to meat-packing plants, led to serious local overgrazing. Several years of continuous drought led ultimately to the death of most of the cattle and many of the owners, as well as total denudation of the landscape, with subsequent erosion.

The volume of food exports, on a world scale, increased about twentyfold between 1854 and 1964 (Grigg 1974). Wheat, barley, maize, and sorghum, important crops of both drought-prone and irrigated areas, have been major components of this increase. The increasing reliance of a largely urban population on a diminishing rural population for food production will ultimately produce great pressure to produce stable quantities of plant products from all agricultural zones, even from zones that are climatically unstable and drought-prone. These pressures are likely to lead in some situations to mismanagement of the environment under conditions of forced production. On the other hand, the same pressures may lead to better selection of drought-adapted crops for these zones. As transport has become cheaper and better organized, humanity has been able to move away from its age-old close association with specific species of plants adapted to local conditions. This association was the basis of subsistence farming. There is now a situation where the modern urban dweller may live in the arctic or equatorial zone and consume plant products from all over the world.

SOCIAL AND ECONOMIC REASONS FOR RESEARCH ON WATER STRESS IN PLANTS

There is a relatively simple scientific reason for conducting research on water stress in plants. Levitt (1978), in a discussion of the terminology associated with water stress, has pointed out that environmental stress has two distinct components—moderate and severe. Moderate stresses have the potential for inducing reversible growth inhibition, whereas more serious stresses have the potential for inducing irreversible cell injury in plants. Thus environmental conditions for plant growth are only optimal when there is no stress. Since all conditions in the environment are probably never simultaneously optimal, plants in nature are almost always being exposed to one or more stresses. Levitt pointed out that the stresses that are of great practical significance in reducing crop yields are the moderate growth-inhibiting stresses, particularly water stress, which can be offset by irrigation. It is only recently, as irrigation has been seen to have a limited potential, that attention has transferred to the more severe cell-injury stresses in the hope of finding ways of limiting these stresses. Clearly, the study of both moderate and severe water stress provides information that can help to optimize plant growth under agricultural conditions. However, there are other compelling, urgent social and economic reasons for trying to optimize crop plant growth in environments exposed to periodic drought.

Desertification is a dynamic process that reduces the productivity of land-use systems in the semiarid and arid zones of the world, where in many countries poverty, underdevelopment, and unemployment exist. Aside from the physical and ecological processes, humanity is the principal agent responsible for desertification (Garduño 1977) through the faulty application of technology. The irrational management of water, soil, and plant resources leads to undernourishment of animals and men, lowered productivity, and ecological deterioration. Garduño points out that any action plan to control desertification must give high priority to preventive techniques since reclamation involves considerable investment, energy inputs, and technical assistance. He also points out that afforestation, revegetation, pasture and crop rotation, and the use of drought-resistant plants are significant factors in increasing productivity, checking desertification, and making the process reversible.

About 60% of the cultivable land of the globe was already under cultivation by 1974 (President's Science Advisory Committee 1974). In the most densely populated countries about three-quarters of the world's population is occupying about one-half of the global cultivable land, of which almost 100% is already under cultivation. In these areas the pressures to cultivate marginal or uncultivable land, with the risk of permanent degradation of the soils, are extremely high. The risk is highest in the arid and semiarid regions, which normally have low population densities more or less compatible with the potentially low biological productivity. In the remainder of the globe (which would include central Africa and South America), where only about one-fifth of the cultivable land is used, increasing the cultivable land means destruction of forests with similar disastrous consequences for soil degradation and erosion. Thus global population pressure has reached the point where pressure on the land has become extreme in some cases, in order to supply food and other agricultural products to indigenous populations or for export to more densely populated urban or industrialized societies.

Since 1650 the increase in world population from about 500 million to the present 4 billion has reduced available crop land from 3 to about 0.4 hectares per capita. Thus in the world today available arable land is not sufficient (even assuming that the energy resources and other technologies were available) to feed the global population a diet similar to that currently consumed in North America or western Europe (Pimentel and Pimentel 1977). On the basis of existing knowledge and agricultural practice, by the year 2000—when the global population is projected to be 6 to 7 billion—only 0.2 hectares will be available per person and producing food will be even more difficult. The need to develop and manage wisely the semiarid and arid zones of the globe for increased crop production demands better knowledge about the nature of water stress and of ways of mitigating its harmful effects on crop production.

Another increasingly important factor globally is the relative scarcity of fresh water and its increasing economic cost. In most regions of the globe, water is usually a significant rate-limiting factor in maintaining high productivity of crop plants throughout the potential growing season. In most circumstances the natural rainfall limits the amount of growth or number of crops that can be grown in a year. Storing rainfall and tapping underground reservoirs of water have traditionally been important ways of increasing crop productivity in many areas of the world, but usually at great cost in terms of human activity, energy inputs, and capital investment. The immense scale leads to enormous investments for some of the giant schemes for creating water reservoirs, such as the Colorado river system in the United States, the Kariba, Volta, Kainji, and Nasser lakes in Africa, and the many schemes in India and the Middle East. A single dam on the river Tigris in Iraq to store 10 billion tons of water for irrigating 400,000 hectares can cost $10 billion dollars (Anonymous 1975). The cost in human labor and capital of the development and maintenance of rice paddies by the hand method in Asia and Southeast Asia over the last several thousand years is incalculable and immense. Nevertheless, it is an indication of the value of conserving water to offset drought periods and stabilize crop production.

The rapidly increasing global population has begun to place serious pressures on the availability and cost of fresh water. The small proportion of global water fixed at any one point in time in biological tissues, primarily plants, is illustrated in Fig. 9. While groundwater reserves appear considerable, they are largely unavailable and generally very slowly released. The high rate of flux of liquid water between rivers, lakes, soil, and the atmosphere, either directly or indirectly through biological organisms, provides a fairly efficient recycling of the limited fresh water. The efficient and economic use of water in agricultural practice is becoming a prime factor in agricultural management. Water-use efficiency can be defined as

$$\frac{\text{crop yield}}{\text{evapotranspiration of crop area}}$$

measured in such units as kg per hectare-cm of evapotranspiration, or kg of dry weight/kg of evapotranspiration. The numerator, crop yield, can be changed appreciably by management, whereas the denominator is largely influenced by climatic conditions. In general, agricultural practice aims at maintaining soil moisture levels at close to field capacity (for example, by drainage or irrigation) since this favors optimal plant growth. Nevertheless, the most efficient water use may be at somewhat lower levels of soil moisture (Doss and Ensminger 1958) which involves some degree of water stress, particularly during less sensitive stages of growth. The most efficient users of water tend to be mesophytic plants, which effectively recover and transpire large volumes of water per unit of land area, but they

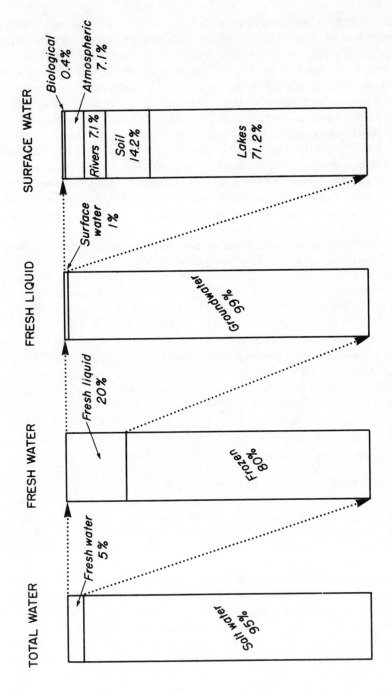

Fig. 9.   The relative availability of water globally.

are not very drought-resistant. The more drought-resistant plants tend to have modest water requirements, produce less dry-matter accumulation per unit area of land, and are better adapted to survival than to high productivity. They are, however, low in efficiency of water use. As the cost of water becomes higher, with man competing directly with agriculture for alternative uses of fresh water (industrial, recreational, and general domestic use, all coupled with increased water pollution), the specific choice of crop and the management of the factors related to both of the quantities in the water-use efficiency definition will become more critical. Plant species with high water-use efficiency potential will need to have some degree of heat and drought resistance as crops are pushed, through the use of irrigation, into the marginal semiarid and arid areas where the very high potential evapotranspiration rates tend to favor lower water-use efficiency. Optimizing water-use efficiency and combining this with an appropriate degree of water-stress resistance in specific crops will require more advanced knowledge about drought resistance than currently exists (Krieg et al. 1977).

Energy considerations are rapidly becoming critical factors in the nature of modern agricultural systems. Knowledge of the energetics of agricultural systems helps to determine which changes should be made, and where they can be made, to take advantage of economic change. Planned and unplanned changes in agricultural management may take place as a result of decisions made in powerful, centrally planned public and private institutions, generally urban-based, that can have far-reaching effects on the utilization and even ecological stability of natural and agricultural zones of vegetation. In many parts of the world, fossil energy has been effectively substituted for manpower. Associated with this change, mainly in the richer countries, is a shift from a plant-protein diet to one heavily reliant on animal protein (Pimentel and Pimentel 1977). For example, of the estimated 1,000 kg of grain produced per capita in the United States, only about 60 kg are consumed directly by humans (U.S. Department of Agriculture 1974, 1975) and the livestock population outweighs the human population more than fourfold. This high emphasis on an animal-protein diet has only been possible through the widespread use of fossil-energy inputs directly in the farm production system and in the food processing and transport system. As an example, to get 375 kcal of beef on the table requires 80 times that amount of fossil energy. Plant food for direct human consumption is also very expensive in terms of fossil energy. A 375-kcal portion of maize (455-g can) consumed in the United States requires 11 times the amount of fossil energy to produce, transport, and market. Thus in the rich countries an increase in volume in food production has been obtained at the expense of a finite resource, fossil energy. This energy source is rapidly increasing in price and becoming scarce, particularly in the underdeveloped world which does not have sufficient purchasing power to compete with the rich countries.

The imbalance in access to fossil energy is turning the richer countries into the food-producing countries that have large quantities of cereal grain for export. Thus production pressure on marginal arid and semiarid areas in the richer countries (western and southwestern United States and Canada, Australia, Argentina, Soviet Union) has increased sharply in this century through the use of mechanization and fossil-energy inputs. The inherent danger is that as the cost of fossil energy increases sharply and exporting countries strive to hold markets to prevent loss of income, serious mismanagement may occur in the drier zones. As an example, the High Plains area of Texas, formerly a region considered too dry or saline for crop production, has become an important corn and sorghum growing area, primarily for cattle feed, as a result of the use of irrigation. However, the area is now faced with dramatic losses in productivity due to the depletion of the Ogalala aquifer, the sole source of fresh water for the area. Hydrologists have predicted that the High Plains area will be the first major agricultural area in the world to run out of irrigation water (Krieg et al. 1977). The only alternatives are to return to dryland agriculture, with a dramatic decrease in productivity, or to improve greatly the irrigation water-use efficiency for the crops currently grown. Alternatively, new crops may be introduced that can achieve the same production with less water or with poorer quality water. Each of the alternatives requires a system in which knowledge about water-stress resistance will be a key factor in management.

Finally there is a completely social reason for understanding more about drought-stress effects on the crop plants that are important in the semiarid zones. It has been estimated that at least 500 million people in developing countries suffer from chronic undernourishment and many of these poorest people in the world depend on sorghum and millet, the two cereal species of the tropics with considerable adaptation to drought. Over the last 20 years the ceiling to the yield in both these crops has been raised dramatically through the use of hybrids made possible by the discovery of cytoplasmic sterility-restorer systems, dwarfing, and disease resistance (Hulse et al. 1980). The full utilization of the potential of these crops will depend upon the proper management of water use and an adequate knowledge of the drought-resistant properties of these species. This latter knowledge is only now forthcoming.

TRADITIONAL PLANT HUSBANDRY PRACTICES TO AVOID DROUGHT STRESS

Grasslands

The occupation by humanity of the drier areas of the globe was generally preceded by a nomadic form of animal husbandry. Nomadism still exists in central Asia (Kirghiz, Kurds, and Mongols) and many parts of Africa (Fulani and Masai tribes), generally on the fringe of desert areas where the growth of grasses and shrubby species is over a short season and the water supply for humans and animals is limited. There is generally a specialized relationship between the domesticated animal and the form of the culture. For example, in central Asia and parts of the Sahel area of Africa, the horse provides the basis for the culture, permitting the coverage of large areas. In the more temperate, mountainous areas where movement is more restricted, the sheep, the yak, or the vicuña may provide the basis of the culture. Cattle and goats are more important in the hot zones.

In the past, provided that the human population was restricted, there was usually a balance between the animal population and the native drought-tolerant or drought-avoiding plant species. However, as human population has increased in surrounding zones, many of the semiarid natural grassland zones have been taken over for arable farming. Examples are the extensive plains and prairies of the United States and Canada, the Pampas of Argentina, and more recently the grazing lands of Kazakhstan in the Soviet Union. Under a nomadic grazing system, overgrazing in dry years or around water holes generally led to a replacement of the more productive species by woody or shrubby drought-resistant species. It would seem that since paleolithic times the impact of humanity on these fragile marginal areas has in general been negative. Husbandry was merely a function of population density rather than a conscious practice, and large areas of the Middle East, North Africa, Greece, Italy, and more recently the southwestern United States are examples of land severely eroded and impoverished through uncontrolled grazing.

The significance of the drought-resistant grasses for soil and water conservation in these zones cannot be overestimated. Their ability to increase the water-absorbing and water-holding capacity of soils is of great significance both in preventing desertification and in reclaiming dry areas. Penetration of rainfall into a silty clay–loam soil under grass can be as much as 7 times as fast as into bare soil (Duley 1952). Similarly, the conservation of runoff which permits springs and rivers to remain active can be directly related to the presence of a grass-covered soil (Semple 1970).

It was not until this century that active steps were taken to reduce the population density of animals and to restrict grazing by the use of fences on these most marginal of dry lands. Prevention of fire has also become important in conservation. Until the advent of the gasoline-powered tractor and the airplane it was not possible or economic to introduce or reestablish either exotic or native forage species on such very extensive areas with low potential productivity.

In the more productive semiarid zones a sedentary grazing system evolved, in association with reliance on crops, to supplement the natural productivity of the original and improved grassland areas. It is in these zones that significant advances in productivity have been made in this century, generally through the introduction, adaptation, and breeding of drought-resistant species with higher forage production potential than the natural species. Examples are the development in Australia of Wimmera ryegrass (*Lolium rigidum*) from introductions from the Mediterranean and the introduction of subterranean clover (*Trifolium subterraneum*) to southern Australia (Harlan 1956). Also, the introduction of crested wheat grass (*Agropyron cristatum*) (Knowles and Buglass 1966) and the breeding of a new creeping-rooted drought-resistant variety of alfalfa (*Medicago sativa* var. Rambler) (Goplen et al. 1980) in Canada have increased the productivity of the northern prairie zones of the United States and Canada.

In summary, then, in the marginal areas of the semiarid zones where grazing of plants is the dominant, or only, form of plant utilization, little has been done to improve drought-stress resistance of plants other than to introduce and oversow adapted drought-resistant species from other areas to replace the local species lost through overgrazing.

Rainfed Cropping in the Dry Zones

Throughout the temperate-to-tropical semiarid zone where annual rainfall is sufficient and reliable, arable cropping has replaced much of the native vegetation. Even where marginal lands were traditionally left for grazing, population pressure, combined with the modern tractor and mechanical plow, have made steady inroads into these areas with, in many cases, increasing frequency of soil erosion and crop failures.

In most cases the farming system is adapted to the often extremely variable rainfall to ensure that the crop is timed to coincide with the maximum availability of water. There is usually some system of adaptation to the occasional good year or disastrous year in terms of precipitation. Frequently, a system of fallow, in alternate years, is used to conserve sufficient soil moisture to provide for a crop in one season. The annual cereal grasses predominate in these systems of dry-land farming, with wheat, barley, millet, sorghum, and the small-grain legumes ranking high in importance. In many of these areas, the rainfall occurs predominantly in the winter months and a rainfall of less than about 250 to 300 mm limits arable farming. In areas of summer rainfall, the limit for arable farming is about 500 mm because of high summer evapotranspiration.

Until the introduction of the mechanical plow pulled either by heavy draught animals or tractors, much of the subsistence farming in these zones—for example, the Middle East and Africa— was carried out on light soils, which could be easily cultivated with hoes or a simple wooden plow. Since the water-retaining capacity of these soils was low, crop yields were also low. With the development of more powerful traction systems, heavier soils and sod-bound soils (such as the temperate grassland prairies of North America and the Argentine) have come under the plow. The heavy clay soils with good water-holding capacity are also extremely difficult to cultivate when dry. The energy input to produce a crop on these heavy soils often exceeds the calorific value of the crop. The generally low yields combined with an erratic climate make these areas economically sensitive to increased fuel prices, and there is now a tendency to increase individual farm size to combat rising costs with economies of scale.

In the Middle East and North Africa the traditional agriculture since Roman times has been almost exclusively based on a winter cereal monoculture. Wheat, valuable in human nutrition, is found in the areas of more reliable rainfall. Barley is grown in the areas of lowest rainfall, colder alpine zones, or semisaline soils. These crops are sown generally after fallow. The period of fallow is longer where rainfall is lowest.

In the Sahel zone of Africa the cereal–fallow system utilizes millets or sorghum (Doggett 1970). Both species have a considerable range of variants adapted to different altitudinal zones and local uses such as human feed, forage, fuel, and making beer. Toward the equator, legumes are often combined in a form of rotation, but the system is basically a crop followed by fallow.

South Africa and South and West Australia have adopted essentially similar systems of rotating crops with pasture. The pasture, which frequently contains a legume such as alfalfa, is used for 3 to 7 years and is followed by a wheat or barley crop for 1 year. In the driest areas the system may be reduced to a wheat, fallow, annual legume in a 3-year rotating system.

In the North American semiarid prairie zone, winter cereals are grown in the south. North of about latitude 48° N, the winters are too severe for survival of winter cereals. In this northern zone, spring-sown varieties of wheat, barley, rye, oats, flax, and rapeseed are the principal crops adapted to the short growing season. Throughout the zone, on the poorer soils, pasture forms part of the long-term rotational system. Forage legumes such as alfalfa, together with the forage grasses, are important in maintaining soil structure. Throughout the prairie region of North America, fallow in every second year is still common.

In South America the dry central Pampas area of the Argentine, previously a natural grassland, has become a dry-land arable farming system, utilizing sorghum, wheat, and other small grains rotated with combinations of grasses and forage legumes such as alfalfa. Sheep utilize the pasture in the colder southern part and cattle in the north. Since there is little surface water, humans and animals rely on deep bores for drinking water. Further east, although rainfall is higher, the very hot summers and strong winds over flat plains give high evapotranspiration rates, and there is commonly a 3-year period of small-grain cereal cropping followed by 6 to 10 years of pasture, predominantly alfalfa.

The roughly similar zones of North and South America, Australia, Africa, the Middle East, southern Soviet Union, and northern China, which have low precipitation, high evapotranspiration, and a short growing season, were all once in natural grassland. Plant production was utilized primarily by grazing with livestock. Increased productivity has been gained by establishing crops suited to direct human consumption. The drought- and heat-resisting small cereal grains such as wheat, barley, and sorghum, which complete their life cycle quickly on the limited available soil moisture, have become widespread in these areas of dry-land farming. Fallowing the soil for one or two years is important in accumulating sufficient soil moisture to ensure a viable crop. The combination of a grass and a forage legume, in rotation with the cereal, counteracts to some extent the tendency to soil erosion by wind and water and in some zones restores nitrogen and organic matter and moderates the process of salinization. Periodic droughts leading to crop failure have significant economic effects in these areas; so, wherever water is available for irrigation and the soils are suitable, irrigation will be practiced to help stabilize production.

Drought stress is thus avoided in these semiarid zones primarily by replacing the native vegetation with introduced specialized crops. Through timeliness of planting, crops can complete their life cycle on the limited precipitation and available soil moisture supplemented by the accumulation of a small reserve from a year of fallow. Two strategies can be adopted for plant breeding in these zones:

1. Develop drought-avoiding cereals which complete their life cycles rapidly, thus ensuring a crop is matured before soil moisture is used up.
2. Develop drought-tolerating cereals which can withstand short intervals of drought without injury and which can utilize any additional rainfall to give increased yields (Kaul and Crowle 1974).

The existensive use of cultivation during the fallow period to break surface crusting, control weeds, and create a dust mulch to reduce surface salinity or reduce moisture loss from compacted soil has been the principal factor in the long-term decline in productivity of many of the semiarid cropping zones. Erosion by heavy rain and surface winds and the salinization of these soils are the direct consequences of cultivation of what was actually a permanent grassland association.

Irrigation

Irrigation is undoubtedly one of the most ancient practical approaches to the prevention of drought stress in plants. The development of the first civilizations in the arid and semiarid zones is closely linked with the development of water utilization and conservation along the margins of principal rivers such as the Nile, Euphrates, Tigris, Indus, and Yangtze (Arnon 1972). Many of the most ambitious engineering projects devised by man have been associated with the conservation, distribution, and application of water to crops, supplementing the deficiency in rainfall in many dry parts of the world.

Irrigation practice seems to have originated independently in China, India, the Middle East, North Africa, and Mexico and Peru. The forerunner to irrigation was probably the occupation of river deltas that flooded annually, such as the Nile, which in its lowest reaches passes through an arid desert zone with little natural vegetation. The construction of barriers to delay the passage of water in streams and small valleys in hilly areas with sporadic rainfall was probably the forerunner of more complex irrigation systems in the arid and semiarid zones. Later, canals were used to tap the flow of rivers to irrigate large areas of land in Asia, North Africa, and more recently Meso-America. Sometimes these canals have reached hundreds of kilometers in length from the original source of the water, and abandoned canals provide the archaeological evidence for the presence of previous civilizations that for various reasons passed away. Throughout the dry zones of western Asia and North Africa, wheat, barley, millets, sorghum, vegetables, and cotton were important crops for irrigation. By contrast, wet-rice cultivation has become one of the distinctive forms of agriculture of eastern Asia in areas that are either sub-humid or humid but, nevertheless, where without irrigation or water conservation plants would suffer drought stress at some point in the growing season (Grigg 1974).

Rice is an important crop that supports much of the rural population of the Far East, which includes the densely populated areas of China, South Korea, Japan, Taiwan, North Vietnam, Thailand, Burma, Bangladesh, the Ganges plain of India, Sri Lanka, and Java. Rice is the dietary staple for more than 60% of the world's people, and about 80% of the world rice crop is grown in Asia and India (Richardson and Stubbs 1978). Upland rice is grown on terraces, formed on sloping, often hilly land in the drier areas, and the crop is brought to maturity solely on rainfall. On the other hand, wet-rice, unlike upland rice which is grown like other cereals, must be submerged beneath slowly moving water to an average height of 100 to 150 mm for three-quarters of the growing season. Thus wet-rice is normally grown in small leveled fields surrounded by low earthen walls, or bunds, which keep the water in. The bunds are later breached to drain the field before harvest. For the above reasons, most wet-rice cultivation is found in deltas or in the lower reaches of rivers where there is little cost in leveling land. Wet-rice farming thus employs a form of water conservation and irrigation that ensures both protection from drought and from temperature extremes. Wet-rice is otherwise well adapted to the high mean monthly temperatures of at least 20°C for 3 or 4 months. About one-half of the wet-rice is grown on conserved flood water or rainfall; the remainder, principally in East

Asia, is irrigated in addition. Wherever wet-rice is grown, it tends to be the dominant and often the only crop.

Until the 1960s there was little change in the pattern of rice farming, which was largely a subsistence form of agriculture on small holdings. However, with the improvement of the production potential by the adoption of hybrid and short-stature varieties in Japan and in the Philippines and through the research and development of the International Rice Research Institute (1976) in the Philippines, there has been an increase in the prevalence of double-cropping with a higher risk of drought stress in this crop. Drought resistance in rice, present in both the upland and lowland varieties, is being improved. Research at the Institute indicates that both types have mechanisms of tolerance to and avoidance of drought at various stages of the life cycle. These attributes are essential since the probability of drought, associated with the beginning and ending of the monsoon, is relatively high in most of the rice-growing areas. Floating varieties of rice have fairly good levels of drought resistance.

Other Cultural Factors That Reduce Drought Stress

Most cultural factors directed at increasing water-use efficiency have some influence on the reduction of drought stress in plants. Thus, wherever arable crops are grown, controlling the density of plant populations is an important practice to match transpiration potential to the seasonal availability of soil moisture. The latter, in turn, is largely determined by precipitation and evaporation. Whenever light is not the principal rate-limiting factor to growth, the degree of control of leaf-area index to optimize soil moisture utilization becomes the major determinant of crop yield. The investigation of optimum row spacing and density of plants within rows has been the subject of a vast amount of agronomic research, both formal and informal. Methods of controlling planting pattern and density range from hand-broadcasting seed to highly sophisticated mechanical methods that space individual seeds accurately in rows, within rows, and at uniform depth in the soil. The latter include such methods as accurately spaced seeds on biodegradable tapes planted in the soil, the vacuum-operated lettuce seed planter, and, recently, machine-extruded gels containing seeds. The latter protect seedlings in the early germination stage from stress and simultaneously provide nutrients, thus contributing to optimum water-use efficiency.

Under some situations, narrowing row spacings may mean reduction in weed competition and less soil crusting from rainfall and hence less direct evaporation of water from the soil surface (Dungan et al. 1958). On the other hand, widening row spacing in very dry situations can optimize the root distribution of forage crops so that competition for limited soil moisture is minimized (Kilcher 1961). As an example, the use of either creeping-rooted alfalfa (Goplen et al. 1980), which can spread among tufted grasses, or the use of alternate rows of alfalfa and grass ensures optimum moisture utilization throughout the soil profile; the grasses utilize the surface moisture and the alfalfa penetrates to the deeper subsoil moisture (Kilcher and Heinrichs 1958).

Controlling planting date is a very old method of utilizing particular species, often with short growth cycles, to ensure crop maturation before soil moisture reserves are expended (Blum 1972). Even in areas where cultivation and sowing is difficult without the onset of rains—for example, in the monsoon areas—yields have been substantially increased by the use of more timely planting (Government of India 1974). One of the most significant benefits from the application of tractor power is improved timeliness of sowing.

Weed control increases the efficiency of water use in a crop. Weeds compete directly with crop plants for soil moisture and nutrients. In addition, most weeds are less efficient than crop plants in the production of dry matter per unit of water transpired. For these reasons, weed control—for example, with herbicides—can substantially increase crop yields and even ensure a crop where maturation might otherwise be difficult due to drought stress. Traditional methods of interrow double cropping prevalent in subsistence farming on small holdings in Africa and Asia can be a practical means of weed control, permitting efficient use of limited seasonal availability of water and crop diversification.

In the early part of this century, tillage was considered to be a significant factor in crop production, partly as a result of the development and application of the gasoline-powered tractor, which substantially increased the speed and ease of cultivation. "Dust mulching" was considered to be an integral part of dry-land farming, at least in the semiarid zone of North America. In fact this excessive cultivation, supposedly to break the crust of fallow soils to improve infiltration, prevent runoff, and simultaneously prevent evaporation by the formation of a dry mulch, played a significant role in wind erosion and, on certain soils, salinization. The latter occurred when salts were brought to the surface by evaporating soil water. The "dust bowl" conditions produced in the 1930s from a combination of the above practice and a series of drought years was thus largely the consequence of mismanagement. More recently, the emphasis has shifted in the opposite direction toward minimizing tillage operations. The benefits from both the absence of tillage and the use of tillage can be combined in such techniques variously described as wheeltrack, plow-plant, and strip tillage where tillage between planted rows is reduced to a minimum (Wiese and Arny 1958). In this system, the higher infiltration of water and the lower soil loss than occurs with conventional tillage systems favors promoted yields (Anderson 1975).

2

A. J. KARAMANOS

The Development of Water Deficits in Plants

Plant water status is determined by the rate of exchange of water between soil and atmosphere through plants. This exchange is expressed by the water-balance equation, which summarizes the gains and losses of water in the root zone of a natural or agricultural ecosystem for a given time interval. Thus,

$$\Delta W = (P + I) - (O + U + ET) \tag{1}$$

where ΔW represents the change in soil water content, $P + I$ the gains (P: precipitation; I: irrigation), and $O + U + ET$ the losses of water from the soil (O: surface runoff; U: deep percolation; ET: evapotranspiration) for the given time interval. When considering the plant as the reference, then soil water constitutes its immediate water supply and the atmosphere the main cause of water loss via evapotranspiration, since the components O and U are of relatively minor importance. Thus, the relative magnitude of water supply and water loss appears to determine the degree to which plants are water-deficient. However, since plants are living organisms, it is important to realize that they do not provide solely a pathway for the movement of water from the soil to the atmosphere. Plants possess mechanisms that effectively control the flow of water according to their water status. These mechanisms, together with the soil and aerial

34   /

environment, provide important parameters in the study of the development of plant water deficits.

In the present chapter the contribution of the aerial environment, the soil, and the plant with its control mechanisms and their importance in the development of plant water deficits will be treated separately.

THE AERIAL ENVIRONMENT

The aerial environment exerts a strong impact on plant water relations. Apart from supplying the ground with rainwater, it also bears two other important features related to the loss of water from the ground. First, it provides the solar energy which induces the evaporation of water; second, it acts as a sink of infinite capacity for the water vapor. In this section the role of the atmosphere in the process of evaporation will be dealt with exclusively.

The Energy Input

The sun and the sky, either directly or indirectly, provide any surface on the ground with radiant energy. Most of this radiant energy lies in the wavelengths between 300 and 3,000 nm (1 nm = 10^{-9} m) and is termed shortwave radiation (S), while the waveband beyond 3,000 nm is known as longwave radiation (L). The latter coincides with the thermal emission by terrestrial surfaces that usually have temperatures around 300°K. Thermal emission strongly depends on the temperature (T) of the emitting body according to Stefan's law:

$$L = \sigma T^4 \tag{2}$$

where σ is the Stefan-Boltzmann constant (5.57×10^{-8} W m^{-2} °K^{-4}). It can be shown that the spectral distribution of the solar radiation coincides with that of a blackbody emitting at a temperature of 6,000°K.

Any natural surface receives and loses amounts of radiant energy. The balance between gains and losses in radiant energy per unit area of a natural surface at a given time is called net radiation (R_n). The gains in energy come through shortwave radiation from the sun and

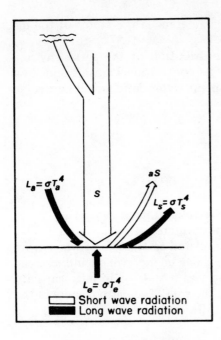

Fig. 10. Diagrammatic representation of the radiation balance of a natural surface. The shortwave radiation (S) includes both direct radiation from the sun as well as diffuse radiation from the clouds. The symbols are the same as in Eq. 3.

sky as well as through the longwave radiation from the surroundings (soil and atmosphere). The losses are the fraction of the shortwave radiation reflected by the surface, together with the longwave radiation emitted by the surface itself (Fig. 10). Thus, R_n is expressed by the energy-balance equation as follows:

$$R_n = (1 - \alpha)S + L_a + L_e - L_s \qquad (3)$$

where α is the reflection coefficient of the surface and L_a, L_e, and L_s are the fluxes of longwave radiation from the atmosphere, the environment (soil and other bodies), and the surface, respectively.

The shortwave radiation term in Eq. 3 exerts an important influence on the R_n of the surface. Thus, the diurnal pattern of R_n is usually determined by the variation in S, since the variation of the longwave components of the energy balance is small or negligible (Fig. 11a). The net radiation R_n takes negative values during the night, when S equals zero and usually $L_s > (L_a + L_e)$. It is also possible that R_n takes negative values during the whole day, especially in winter and in high latitudes (Fig. 11b).

A significant role in the magnitude of R_n is also played by the reflectivity of the surface. The factor α, also known as surface albedo, depends on solar elevation and the physical characteristics (for example, color, roughness) of the surface. For fresh snow, α is maximal (0.8 to 0.9); for clear deep water, it is minimal (0.05 to 0.2). Monteith (1975a) tabulated values of α for a series of plant and animal surfaces; individual leaves usually reflect 28 to 34%, farm crops 15 to 26%, and natural vegetation and forests 12 to 24% of the incident solar radiation. The possibility of changing the net radiation of plant surfaces by altering their reflecting properties looks promising (see, for example, Fuchs et al. 1976).

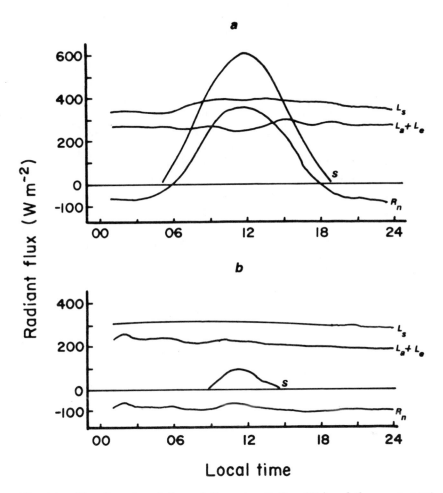

Fig. 11. The diurnal variation of the net radiation (R_n) and the components of the radiation balance at Bergen, Norway (60° N, 5° E). (a) In the spring (April 1968). (b) In the winter (January 1968) where R_n is negative throughout the day. (From *Principles of Environmental Physics* by J. L. Monteith, in the series "Contemporary Biology." Copyright © 1975 Edward Arnold Ltd.)

The net radiation of a natural surface is dissipated by means of sensible-heat exchange with the soil (G) and the atmosphere (H). If the surface is wet, a portion of R_n causes water to evaporate and, accordingly, is converted into latent heat (λE). Finally, if plant surfaces are considered, a part of R_n is also converted into energy of chemical bonds via photosynthesis (aA). In general, the dissipation of net radiation can be written as:

$$R_n = H + G + \lambda E + aA \qquad (4)$$

where the factors λ and a represent, respectively, the latent heat of vaporization of water (about 600 cal g^{-1}) and the chemical energy storage coefficient (about 3,600 cal g^{-1}). They are used to convert the amounts of the evaporated water and the produced dry matter into their energy equivalents. The daily magnitude of the storage term G is usually small, given that inward flux by day usually balances outward flux at night, while the chemical storage term aA seldom accounts for more than 5% of R_n. Thus, in practice the net radiation is thought to be dissipated both as sensible- and latent-heat fluxes. The relative magnitude of these two terms depends on the wetness of the surface, that is, on the availability of water for evaporation. Whenever water is readily available, λE greatly exceeds H (Fig. 12). In soils under vegetation, λE is the energy equivalent of evapotranspiration, which is the main source of water loss from the soil in the water-balance equation (Eq. 1). In agricultural lands, λE usually predominates over H (Milthorpe and Moorby 1974).

Transport of Water Vapor to the Atmosphere

Once the required amount of energy is provided, evaporation starts with the transport of water-vapor molecules from the wet surfaces to the atmosphere. The transport occurs by molecular diffusion and turbulent eddy movement. The former occurs in the vicinity of the evaporating surface, inside the disturbed air-boundary-layer that sheathes all natural surfaces. Beyond the boundary layer there is a transitional region where both molecular diffusion and turbulent eddy movement are involved. Then, at a certain distance away from the surface, transport occurs exclusively by turbulent mechanisms.

It is generally accepted (Monteith 1965) that exchanges of matter between physical surfaces and the atmosphere follow the electrical network analogue (Ohm's law) according to which:

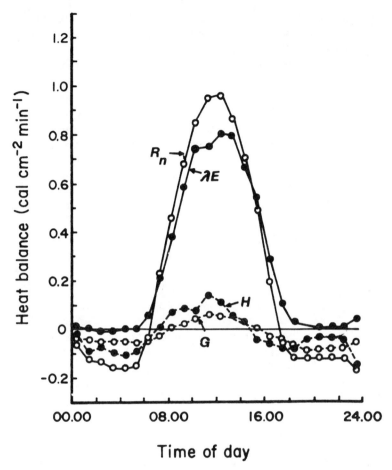

Fig. 12. Diurnal changes in the energy-balance components over pasture during clear summer-day conditions in Wisconsin (after Tanner 1960).

$$\text{Flux} = \frac{\text{potential difference}}{\text{resistance}} \qquad (5)$$

Thus, irrespective of the nature of the transport mechanism, the net transfer of water vapor from the evaporating surfaces to the atmosphere can be written as:

$$E = \frac{c_s - c_a}{r_a} = \frac{273\rho_v}{PT} \frac{e_s - e_a}{r_a} \qquad (6)$$

where c_s and c_a are the vapor concentrations at saturation and in the bulk air (g cm^{-3}) and e_s and e_a the vapor pressures at saturation and in the bulk air (mbar), respectively. r_a is the resistance to the transport of vapor from the evaporating surface to the bulk air (s cm^{-1}). It can be seen that either vapor concentration or vapor-pressure difference can be considered as the driving force for vapor movement. The term $273\rho_v/PT$ (where ρ_v is the density of water vapor in g cm^{-3}, P the atmospheric pressure in mbar, and T the absolute temperature in $^\circ$K) converts vapor concentration into vapor pressure.

When considering soil under vegetation, evaporation occurs in the leaf mesophyll tissue. Then, vapor moves through the stomata and the air boundary layer to the bulk air. Consequently, Eq. 6 becomes:

$$ET = \frac{c_1 - c_a}{r_g + r_a} = \frac{c_1 - c_o}{r_g} = \frac{c_o - c_a}{r_a} \tag{7}$$

In this case c_1, c_o, and c_a are the vapor concentrations at the leaf liquid–air interfaces, at the crop surface, and in the bulk air, respectively; r_g and r_a are the resistances to vapor transport through the leaf and the air boundary layer, respectively.

The magnitude of the numerator of Eqs. 6 and 7 depends on the dryness of the atmosphere (c_a or e_a) as well as on the temperature of the evaporating surface, which determines the value of the saturation vapor pressure; the latter is regarded as identical to the vapor pressure at the evaporating surface (Monteith 1965). It is because of the strong dependence of e_s on temperature (Fig. 13) that energy supply affects evaporation, primarily through changes in e_s or c_s.

The resistance r_a, known as boundary layer or aerodynamic resistance, is a function of the windspeed. The windspeed over a natural surface (u) increases exponentially with height (z) above the surface (Fig. 14). Accordingly, there is a linear relation between $u(z)$ and ln z. The roughness of the surface, which is a function of the height of the roughness elements (h), determines the height at which windspeed becomes zero, called roughness length (z_o). z_o is related to the height of the roughness elements ($z_o = 0.13h$; Monteith 1975a), and therefore z_o increases with increasing roughness of the surface (Fig. 14). The windspeed at a height z, $u(z)$, can be easily derived as:

$$u(z) = a \ln\frac{z}{z_o} \qquad \text{(for a free water surface)} \tag{8}$$

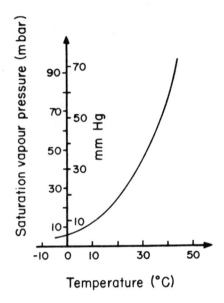

Fig. 13. The saturation vapor pressure of water as a function of temperature.

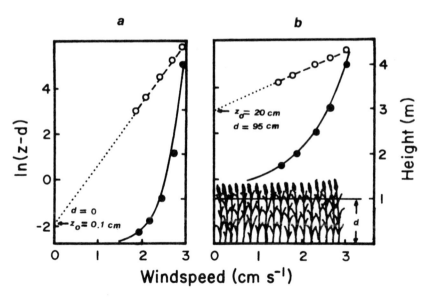

Fig. 14. Wind profiles over a free water surface (a) and a tall crop of a height of about 1.3 m (b) when the windspeed at 4 m above ground is 3 m s^{-1}. The filled circles represent measurements from sets of anemometers while the open ones are the values of ln $(z - d)$ plotted against windspeed in both cases. The various parameters of the aerodynamic resistance (z_o, d) are also indicated. (Adapted from *Principles of Environmental Physics* by J. L. Monteith, in the series "Contemporary Biology." Copyright © 1975 Edward Arnold Ltd.)

$$u(z) = a \ln \frac{z - d}{z_o} \qquad \text{(for a crop surface)} \qquad (9)$$

where a is a constant. The parameter d is called zero-plane displacement and is used whenever a surface with finite roughness elements (crops or surfaces with natural vegetation) is considered. It is generally assumed that $d = 0.63h$ (Monteith 1975a). Thus, Eqs. 8 and 9 suggest that windspeed increases faster with height as the surface becomes smoother (Fig. 14). Monteith (1965) derived the following expression for the aerodynamic resistance:

$$r_a = \frac{[\ln (z - d)/z_o]^2}{uk^2} \qquad (10)$$

where k is von Karman's constant, with a mean value of about 0.41. An example of the influence of windspeed on r_a for different values of z_o is shown in Fig. 15.

The additional resistance (r_g) encountered when evaporation occurs via plant surfaces (Eq. 7) is of great ecological importance and

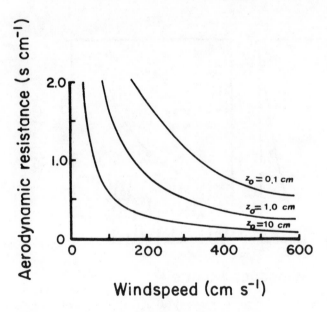

Fig. 15. The dependence of the aerodynamic resistance on the windspeed at various roughness lengths (z_o), calculated from Eq. 10 (after Monteith 1965). (Reproduced by permission of the Society for Experimental Biology.)

greatly affects evapotranspiration. In well-hydrated plants, r_g is very small. Thus, the evapotranspiration from a soil completely covered by vegetation well supplied with water is very close to the evaporation from an open water surface and is called potential evapotranspiration (ET_p). When soil water supply is restricted to a certain degree, plants increase the r_g and then ET falls below the potential rate ET_p. More detailed information on the nature of r_g will be given later.

The rate of evaporation (E) is a very important ecological parameter for a given region because it characterizes the rate at which the atmosphere is capable of extracting water from a completely wet surface. In other words, E expresses the atmospheric evaporative demand and constitutes a meteorological parameter for comparing different regions and different periods within a region. It is also useful for agricultural purposes, especially for irrigation scheduling, since it gives a basis for the calculation of crop water consumption. Among the different formulas suggested for the calculation of the evaporative demand from meteorological data, the one derived by Penman (1948) is the least empirical and the most physically sound. Penman considered evaporation as the resultant of both an energy supply and an aerodynamic vapor transfer from the wet surface. Thus in his formula there is an energetic term (the net radiation received by the surface, R_n) and an aerodynamic term (E_a):

$$E = \frac{\Delta R_n + \gamma E_a}{\Delta + \gamma} \tag{11}$$

where $\Delta = de_s/dT$, that is, the slope of the saturation vapor pressure vs. temperature curve (mbar deg^{-1}; see Fig. 13), and γ is the psychrometric constant (mbar deg^{-1}). Penman considered E_a as a function of both windspeed and saturation deficit: $E_a = f(u) \times [e_s(T_a) - e_a]$. The term $e_s(T_a)$ is the saturation vapor pressure at the air temperature.

Monteith (1965) replaced the aerodynamic term E_a with the expression $\rho_a c_p [e_s(T) - e_a]/\gamma r_a$. Thus the original Penman expression was modified as follows:

$$E = \frac{\Delta R_n + \rho_a c_p [e_s(T) - e_a]/r_a}{\Delta + \gamma} \tag{12}$$

where ρ_a is the density and c_p the specific heat of air (cal g^{-1} deg^{-1}).

For soil surfaces under vegetation, Monteith added another factor containing the plant resistance (r_g) in the denominator of Eq. 12. So, the evapotranspiration rate is given by the expression:

$$ET = \frac{\Delta R_n + \rho_a c_p [e_s(T) - e_a]/r_a}{\Delta + \gamma(1 + r_g/r_a)} \tag{13}$$

By dividing Eqs. 12 and 13 we find:

$$\frac{ET}{E} = \frac{\Delta + \gamma}{\Delta + \gamma(1 + r_g/r_a)} \tag{14}$$

Thus the fraction ET/E depends only on air temperature and the ratio r_g/r_a. This is illustrated in Fig. 16.

Equation 14 and Fig. 16 reconfirm why evapotranspiration is always lower than the evaporation rate from an open water surface ($r_g = 0$). For many field crops, the rate of potential evapotrans-

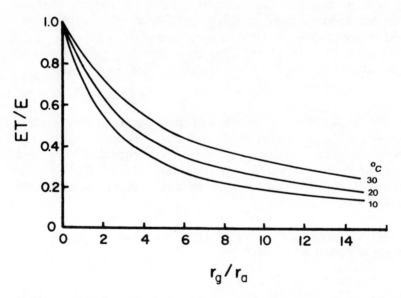

Fig. 16. Changes of the ratio of evapotranspiration over evaporation rate from a free water surface (ET/E) with the variation in the ratio of the plant resistance over the aerodynamic resistance (r_g/r_a) for different temperatures (after Monteith 1965). (Reproduced by permission of the Society for Experimental Biology.)

piration (r_g/r_a between 1 and 2) is about 60 to 80% of the evaporation rate (Monteith 1965).

In summary, then, evaporation rate depends on four meteorological parameters: incident radiation, temperature, humidity, and windspeed. When plant resistance is constant, evapotranspiration increases (1) linearly with radiation, (2) linearly with saturation deficit, and (3) with air temperature when both radiation and saturation deficit are constant. Finally, the variation with windspeed is more complex.

THE SOIL

The soil is the basic source of water for plants. It is important, therefore, to examine the soil properties that determine the availability of water to plants. These properties are related to the forces by which water is retained in the soil. The soil is a heterogeneous system consisting of a solid phase (the soil particles), a liquid phase (the soil solution), and a gaseous phase. Thus, in the most common case of unsaturated soils, the water-retaining forces can be separated into three groups: (1) those arising at the liquid-air interfaces, (2) those arising at the solid–liquid interfaces, and (3) those related to the presence of solutes in the soil solution.

Water-Retaining Forces

Forces at the Liquid–Air Interfaces

The mineral and organic compounds of the soil form a solid matrix consisting of separate particles as well as of aggregates of many particles. The mutual arrangement of particles and aggregates in the soil determines the characteristics of the soil pore space in which water and air are transmitted or retained. In general, soil pores are regarded as being interconnected within a solid phase to form a continuous network of capillary channels of various diameters. Accordingly, in unsaturated soils where air occupies a considerable fraction of the pore space, water is retained by forces of capillarity. In a capillary of radius R, water is retained with a tension τ arising from its surface tension σ (dyne cm^{-1}), according to the formula:

$$\tau = \frac{2\sigma \cos \phi}{R} \tag{15}$$

where ϕ is the curvature of the water–air interface, measured as the contact angle of the water with the walls of the capillary (Fig. 17a). Obviously, as the capillary loses water, the curvature increases (Fig. 17b), the angle ϕ tends to zero, and τ increases greatly. The strong dependence of the tension τ on the radii of the capillaries can be seen in Table 3. The tension τ is identical to the term "suction" used to describe the negative pressure that retains water in porous materials.

Evidently, the size of the pores is extremely important for soil water retention, though the size varies greatly in a given soil. Nevertheless, the distribution of pore sizes shows similarities in soils of similar texture. The term texture refers to the "feel" of the soil material, whether coarse and gritty, or fine and smooth. Quantitatively, soil texture is related to the relative proportions of three basic sizes of soil particles, namely of sand, silt, and clay, in the soil. These three main textural fractions are distinguished on the basis of the diameter of their particles (Table 4) as determined from soil mechanical analysis.

In general, fine-textured soils have smaller pores than coarse and medium soils. However, differences may exist in the range of pore sizes even in soils of similar texture (Fig. 18). Some soils are characterized by a variety of pore sizes, while others contain a preponderance of pores of a certain size. In the former case, the soils lose water more gradually, through a wide range of soil suctions, than in the latter case, where most of the retained water is lost more abruptly.

The great significance of the pore-size distribution on soil water retention is obvious when the curves relating soil water content to soil water suction or the soil water characteristic curves (Fig. 19) are examined. These curves differ markedly among soils of different texture. In general, the water content at a given suction is greater and the slope of the curve less abrupt with increasing clay content in the soil. In a sandy soil, most of the pores are large; once these are emptied at a given suction, only a small amount of water remains. On the other hand, in a clayey soil, pores are much smaller, the pore-size distribution is more uniform, and, accordingly, pores lose water at greater suctions and more gradually.

Soil structure—the mutual arrangement, orientation, and organization of the soil particles—also affects the shape of the soil water

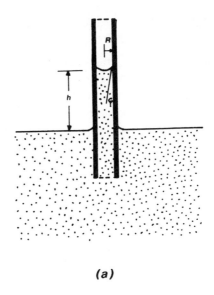

(a)

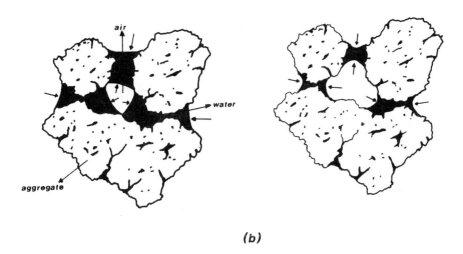

(b)

Fig. 17.   (a) Quantities involved in the capillary rise of a liquid. R is the radius of the capillary, ϕ the curvature, and h the height of the capillary rise. (b) Soil water retained within and between adjacent soil aggregates in two stages of hydration. In the more dehydrated stage (right) the curvature of the water menisci is increased (see points indicated by arrows in both stages).

TABLE 3: Values of Water Tension (τ) for Different Radii (R) of Capillaries at 20°C

(Surface tension of water was taken as 72.8 dyne cm^{-1} and the contact angle as zero)

R (cm)	τ (bar)
10^{-1}	1.46×10^{-3}
10^{-3}	1.46×10^{-1}
10^{-5}	1.46×10
10^{-7}	1.46×10^{3}

characteristic curve by altering pore-size distribution. For example, soil compaction reduces considerably the size of the large pores with little or no effect on the intermediate and micropores, respectively. This causes a variation of the water characteristic curves of a given soil type with depth (Fig. 20). In general, soil compaction makes the water characteristic curves of coarse soils resemble those of fine-textured ones.

Forces at the Liquid–Solid Interfaces

The second mechanism of water retention mainly arises in clayey soils, where swelling and shrinking accompany hydration and dehydration, respectively. These changes in volume are the result of the structure of the clay minerals, the primary components of the clay fraction. The clay minerals are plate-like in form, and their crystalline structure consists of sheets of oxygen and hydroxyl ions

TABLE 4: Textural Classification of Soil Fractions According to Particle Diameter Ranges as Adopted by the International Society of Soil Science

Name of Fraction	Range of Diameter of Particles
Gravel	above 2 mm
Coarse sand	2.0 to 0.2 mm
Fine sand	0.2 to 0.02 mm
Silt	0.02 to 0.002 mm
Clay	below 0.002 mm

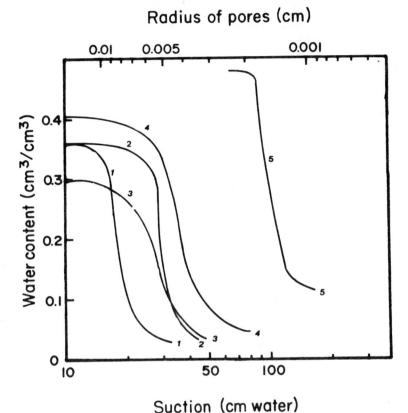

Fig. 18. Relation between soil water content, suction, and pore size in non-swelling soils. Curves 1 to 4 refer to sands with particles of different diameters (1: 1 to 0.5 mm; 2: 0.5 to 0.25 mm; 3: 0.4 to 0.25 mm; 4: find sand) while 5 refers to a slate dust. (From *Relations between Water and Soil* by T. J. Marshall. Copyright © 1959 Commonwealth Agricultural Bureaux.)

with metallic cations (mainly silicon and aluminum) fitting into the spaces and bonded to these layers. The various types of clay minerals are characterized on the basis of the way that silica and alumina sheets are interlinked and stacked together (Fig. 21). In some types of minerals (for example, illite, kaolinite), the chemical bonds between adjacent sheets are strong enough to impart a more-or-less rigid structure to the mineral. In others (for example, montmorillonite, vermiculite), chemical bonds do not exist and water molecules are allowed to enter between the sheets of the mineral, thus expanding the crystal lattice considerably. Therefore, clays with a high content of either montmorillonite or vermiculite exhibit marked volume vari-

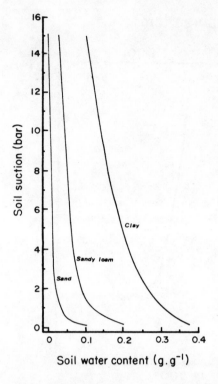

Fig. 19. Soil water characteristic curves for sand, sandy loam, and clay soils. (With permission from *Plant-Water Relationships* by R. O. Slayter, 1967. Copyright by Academic Press Inc. (London) Ltd.)

ations with changes in their water content. Such soils often develop broad cracks during prolonged droughts.

All clay minerals carry a net negative charge, which arises from an incomplete charge compensation of the terminal oxygen atoms at the edges of the mineral sheets as well as from substitution

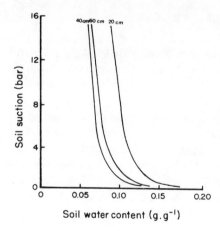

Fig. 20. Soil water characteristic curves of a sandy loam soil at three different depths (20, 40, and 60 cm) (Karamanos, unpublished).

a   b

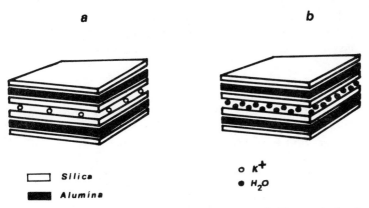

Silica

Alumina

o K^+
● H_2O

Fig. 21. Diagram showing the arrangement of silica and alumina sheets for two types of clay minerals. (a) Illite, nonswelling. Potassium cations enter between adjacent silica sheets and are held by chemical bonds strong enough to prevent separation and swelling. (b) Montmorillonite, swelling. Instead of potassium, water molecules enter between the silica sheets, causing separation and swelling of the clay (redrawn from Kramer 1969). (From *Plant and Soil Water Relationships. A Modern Synthesis* by P. J. Kramer. Copyright © 1969 McGraw-Hill Book Company. Used with permission of McGraw-Hill Book Company.)

of silicon and aluminum by cations of lower valences (Mg^{2+}, Fe^{3+}). The negative charge of the clay particles is neutralized by exchangeable cations (mainly Ca^{2+}, Mg^{2+}, K^+, and Na^+) attracted from the soil solution. Water dipoles are also attracted by the negative charges to form a hydration layer similar to that observed in colloids.

Dehydration in nonshrinking soils is followed by the entrance of air in the pore space and the development of the surface-tension forces described in the previous section. In shrinking soils, however, the reduction in volume after dehydration excludes the entrance of air in the pore space. The exchangeable cations in a water layer between two parallel clay plates increase the osmotic pressure of the solution locally. Ionic concentration and osmotic pressure increase continuously as water is removed and the plates are brought to closer proximity. Therefore, water has an increasing tendency to be redrawn into the layer between the clay plates. The electrostatic repulsive forces that develop as clay plates are brought closer and closer also induce an increasing tendency of the water molecules to separate the plates.

In conclusion, in soils of high clay content, the filling and emptying of pores on wetting and drying may be less important than swelling and shrinking. Thus, in contrast with sandy soils, the shape of the soil water characteristic curve depends less on the pore structure of the soil and more on the size distribution and the properties of the clay particles.

Forces Arising from the Presence of Solutes

The presence of soluble salts in the soil increases the osmotic pressure and lowers the vapor pressure of the soil solution. In contrast with the previously described forces, solutes do not directly influence the amount of water retained against pressure or tension. The osmotic contribution of the soil solution may become important whenever water flow occurs through semipermeable membranes, as, for example, through plant roots. Accordingly, solutes influence the availability of soil water to plants.

Energetic Aspects and Terminology of the Soil Water Status

To study the movement of water within the soil and toward root surfaces, it is important to know how far the water is from equilibrium at different points within the soil; in other words, to know its potential energy in relation to that of water in other soil regions. The differences in the energy status of water between one point and another cause a flow of water within the soil from higher to lower energy levels for the achievement of equilibrium. Buckingham (1907) first introduced the term capillary potential as a criterion for the energy status of soil water. The concept was further elaborated by different investigators (Edlefsen 1941; Childs 1957; Bolt and Frissel 1960) to reach a formal definition by the International Society of Soil Science (Aslyng 1963): the total potential of soil water is "the amount of work that must be done per unit quantity of pure water in order to transport reversibly and isothermally an infinitesimal quantity of water from a pool of pure water at a specified elevation at atmospheric pressure to the soil water (at the point under consideration)." In this definition, the energy status of pure free water is set as an arbitrary measure for the expression of soil water potential. The various force fields acting on soil water cause its energy status to differ from that of pure free water. For example, all forces involved in the retention of soil water tend to lower its potential in comparison with that of pure water. On the other hand, when considering points below the soil water table, the hydrostatic pressure arising from the height of water above the point makes the water potential higher than that of pure free water.

According to Slatyer and Taylor (1960), the energy status of water in a system can be expressed thermodynamically in terms of the chemical potential (μ_w) or the partial molal Gibbs free energy ($\overline{G}_w$) of the water in the system as follows:

$$\mu_w = \overline{G}_w = \left(\frac{\partial G}{\partial n_w}\right)_{T,P,n_s} \tag{16}$$

where G is the Gibbs free energy of the system (a function of its internal energy) and n_w the number of water moles. Thus, μ_w or $\overline{G}_w$ are defined more simply as the change that occurs in the energy content of a system for a given change in the number of moles of water in the system at constant temperature (T), pressure (P), and solute content (n_s). Furthermore, the water potential of a system (Ψ) is defined as:

$$\Psi = \frac{\mu_w - \mu_w^0}{\overline{V}_w} \tag{17}$$

where μ_w^0 is the chemical potential of pure free water and $\overline{V}_w$ the partial molal volume of water (18 cm^3 g^{-1}). As defined, the water potential provides a measure of the capacity of the water at a point in the system to do work in comparison with the work capacity of pure free water (μ_w^0), which is taken arbitrarily as zero. Ψ is expressed as energy per unit volume. This is dimensionally equivalent to pressure, and thus traditional pressure units (1 bar, 1 atm, 1 MPa, etc.) are used for the expression of Ψ. The concept of water potential provides a unified measure for the energy status of water at any place within the soil-plant-atmosphere continuum.

Soil water potential is thought to consist of component potentials that represent the different kinds of forces involved in soil water relations.

$$\Psi_{\text{soil}} = \psi_m + \psi_s + \psi_p \tag{18}$$

where ψ_m, ψ_s, and ψ_p are the matric, solute, and pressure potentials, respectively. The *matric potential* (ψ_m) refers to the forces arising both at the liquid–air and solid–liquid interfaces. In sandy soils, ψ_m is mainly determined by forces of capillarity, while forces of mutual repulsion of soil particles also play a significant role as the clay fraction of the soil increases. Both kinds of forces attract and bind water in the soil and accordingly reduce its potential energy below that of pure water. Therefore, ψ_m takes negative values. ψ_m is identical to the capillary potential introduced by Buckingham (1907). The *solute* or *osmotic potential* (ψ_s) refers to the forces arising from the presence of osmotically active substances in the soil. These solutes lower the vapor pressure and hence the potential energy of soil water. Thus ψ_s also takes values below zero which are proportional to the concentration of osmotic substances (C_s) in the soil solution:

$$\psi_s = -RTC_s = -\pi \tag{19}$$

where π is the osmotic pressure of the soil solution. The *pressure potential* (ψ_p) develops whenever a point in the soil is under a pressure greater than the local atmosphere. Such situations are usual when considering points below the water table in flooded or saturated soils. There, ψ_p is equal to the hydrostatic pressure exerted by the height of the water table above the point under consideration (h):

$$\psi_p = \rho_w gh \tag{20}$$

where ρ_w is the density of water and g is the acceleration due to gravity. Obviously, ψ_p takes only positive values.

In normal agricultural practice, the layer of soil accessible to plant roots is unsaturated for the greatest part of the growing season; therefore, ψ_p equals zero. In soils low in salt, ψ_s is also zero or negligible, and thus soil water potential is essentially equal to ψ_m. In such situations, measurements of ψ_m (for example, by using tensiometers) give an adequate estimate of Ψ_{soil}.

The component potentials mentioned above refer to force fields arising within the soil mass. However, external force fields also affect the energy status of water in soils. Among them, the gravitational field is the most important and so an additional component potential, the *gravitational potential* (ψ_g), can be added to Eq. 18 to give the total *soil water potential* (Φ):

$$\Phi = \Psi_{soil} + \psi_g = \psi_p + \psi_s + \psi_m + \psi_g \tag{21}$$

At a given point, ψ_g is determined by the elevation of the point above some arbitrary reference level. The reference level is usually set at a point below the considered soil profile, so that ψ_g can be taken as positive or zero. At a height z above the reference level, the gravitational potential (potential energy) per unit mass of water is:

$$\psi_g = gz \tag{22}$$

Water content

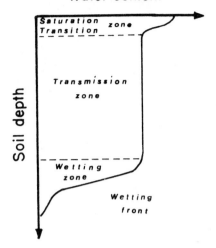

Fig. 22. Typical pattern of a moisture profile during infiltration. (With permission from *Agricultural Physics* by C. W. Rose. Copyright © 1966 Pergamon Press Ltd.)

while the expression per unit volume becomes:

$$\psi_g = \rho_w g z \tag{23}$$

ψ_g is usually neglected when studying horizontal water movement, as, for example, water movement toward plant roots. On the other hand, ψ_g is important in studies of vertical water movement within the soil profile. In saturated and wet soils, the total water potential Φ is mainly a function of ψ_g because the other component potentials are nearly zero (Eq. 19). However, as soon as the soil becomes unsaturated, ψ_m decreases rapidly and the relative importance of ψ_g declines progressively.

Movement of Water in Soils

Water Entry into the Soil

When water is applied to the soil surface by rain or irrigation, it moves downward through a process known as infiltration. The distribution of infiltrating water with depth in the soil, referred to as the moisture profile, follows a general path first described by Bodman and Coleman (1944) (Fig. 22).

There is a restricted saturation and transition zone just below the soil surface followed by a transmission zone of almost constant

Soil water content (cm³ cm⁻³)

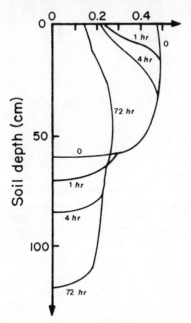

Fig. 23. Moisture profiles during the redistribution of soil water after infiltration into a slate dust. The times by the curves are measured from the moment of cessation of infiltration (after Youngs 1958). (Reproduced by permission of the Society for Experimental Biology.)

water content. Water moves from the transmission zone to the wetting zone where water content changes markedly with both depth and time. At the bottom of the wetting zone there is a sharply defined wetting front, the boundary between the wet and dry soil. The moisture profile changes with time in a manner predicted by Philip (1957) (Fig. 23). The transmission zone deepens continuously with a concomitant downward movement of the wetting zone and front. The latter becomes less steep as it moves deeper.

The rate by which the added water is absorbed by the soil, the *infiltration rate*, is an important soil parameter in studies of soil and water management. In relation to the rate of water supply, it determines the amount of water entering the soil at a given instance. When the rate of water supply through rain or irrigation exceeds the soil infiltration rate, water starts to run off on the soil surface. The surface runoff of water (Eq. 1) is undesirable in agriculture both in terms of water economy and soil erosion. The infiltration rate decreases with time after the application of water on the surface, so that a fairly stable minimum value is reached after a certain time (Fig. 24).

Philip (1957) proposed the following expression, which was based on a theoretical treatment of the process:

$$i = st^{1/2} + wt \qquad (24)$$

where i is the cumulative infiltration (cm^3 cm^{-2}) at time t (s) and s and w are constants having a physical meaning. The term $st^{1/2}$ describes the contribution of capillarity, s being a measure of soil water uptake as related to initial water content, while the term wt refers to the contribution arising from gravity. The presence of the constant s in Eq. 24 implies a dependence of the infiltration rate on the initial soil water content. When this is high, a great proportion of

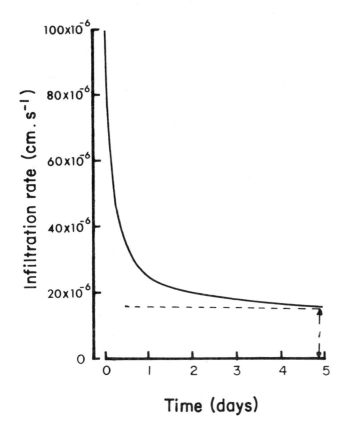

Fig. 24.   The pattern of the rate of infiltration against time for an initial soil water content of 0.24 cm^3 cm^{-3} soil; i denotes the final infiltration rate (adapted from Philip 1957).

TABLE 5: Minimum Infiltration Rates of Various Soils Arranged in Four Groups

Types of Soil	Rate (cm hr^{-1})
Deep sands, deep loess, aggregated silts	0.75 to 1.125
Shallow loess, sandy loams	0.375 to 0.75
Many clay loams, shallow sandy loams, soils low in organic matter, soils high in clay	0.125 to 0.375
Soils of high-swelling, sodic soils	below 0.125

Source: After Musgrave (1955).

pores is already filled with water and hence the rate of water absorption is reduced. Soil texture and structure exert an important influence on infiltration rate. Musgrave (1955) distinguished four groups of soils in terms of their minimum infiltration rates (Table 5).

It appears that higher final infiltration rates are associated with coarse-textured soils where large pores prevail. The values cited in Table 5 indicate a considerable range of infiltration rates. The existence of zones of low permeability, such as surface crusts, clay and cultivator pans, etc., further reduce infiltration rates even in light soils. The type of vegetation, the management practice, and the way in which water is applied are some other factors affecting infiltration rates.

Movement of Water within Soils

Liquid Water Movement—General Aspects Liquid water movement occurs along gradients of total soil water potential ($d\Phi/dz$). Usually, soil water movement is treated separately in saturated and unsaturated soils.

In saturated soils, all pores are filled with water, and thus the driving force $d\Phi/dz$ coincides with $d\psi_g/dz$ because gravity largely determines soil water movement. The theory of saturated water flow is based on Darcy's law, which states that water-flux density (q, the volume of water passing through a unit cross-sectional area per unit time) is proportional to the driving force:

$$q = -K \frac{d\Phi}{dz} \tag{25}$$

where K is the hydraulic conductivity (cm s^{-1}). Since q is also expressed in cm s^{-1}, Φ is in units of equivalent cm of water (1 bar = 1,017 cm of water). The negative sign denotes flux from higher to lower potential levels. The hydraulic conductivity for water is an important soil characteristic that varies enormously in different soil types (from less than 3 x 10^{-5} to more than 7 x 10^{-3} cm s^{-1}; see O'Neal 1952). K is primarily influenced by the pore-size distribution of a given soil but also by factors affecting soil porosity such as the degree of swelling of the soil colloids, the concentration of electrolytes in the soil solution, and the management history.

In unsaturated soils Darcy's law is still valid, but with some modifications. First, the gravitational potential rapidly becomes less important as air enters the soil pores and matric potential begins to dominate. Thus the driving force becomes $d(\psi_g + \psi_m)/dz$. Second, hydraulic conductivity cannot be regarded as a constant any more, since it is also a function of soil water content (Childs and Collis-George 1950). K is found to decrease rapidly with falling soil water content (Fig. 25). The reason for such a behavior of K can be visualized when considering various stages of dehydration of an aggregated soil (Fig. 26).

At saturation, both macropores (the channels formed between the aggregates) and micropores (the microchannels within aggregates mainly determined by soil texture) are filled with water (Fig. 26a). As air enters the soil, liquid continuity in the macropores is broken and water moves near the points of contact of the aggregates (Fig. 26b), while at the same time the cross-sectional area available for water flow decreases. In a more dehydrated stage (Fig. 26c), the micropores themselves begin to dehydrate. This causes both a much greater resistance to water flow and a further reduced area for water flow.

When considering a given soil layer of thickness z, the rate of change in soil water content in that layer ($d\theta/dt$) can be regarded as equal to the rate of water flux into and out of the soil layer (dq/dz), according to the continuity equation

$$\frac{d\theta}{dt} = -\frac{dq}{dz} \qquad (26)$$

or (see Eq. 25)

$$\frac{d\theta}{dt} = \frac{d}{dz}\left(K\frac{d\Phi}{dz}\right) \qquad (27)$$

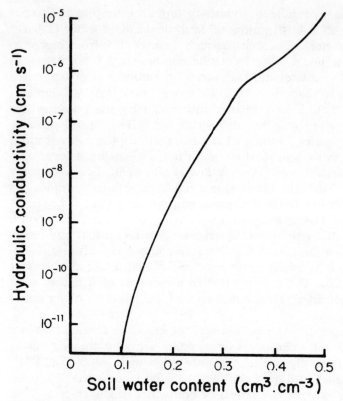

Fig. 25. The relation between the hydraulic conductivity and water content for Yolo light clay soil (after Philip 1957).

The total potential Φ can be considered as the sum of only two components, namely ψ_m and ψ_g. Since the component potentials must be expressed in cm of equivalent height of water, then $\psi_g = z$ and thus $\Phi = \psi_m + z$. Then, Eq. 27 becomes:

$$\frac{d\theta}{dt} = \frac{d}{dz}\left(K\frac{d\psi_m}{dz}\right) + \frac{dK}{dz} \tag{28}$$

The first term on the right-hand side accounts for the effects of capillarity, while the second accounts for the effect of gravity. Obviously, if the flow is horizontal, the second term can be eliminated.

Redistribution Patterns of Liquid Water within the Soil Profile The patterns of water distribution in the soil caused by factors other than root action will be briefly examined here. Patterns arising from root activity will be dealt with later.

After the cessation of infiltration with the soil surface maintained at saturation, the water moves downward in a pattern already shown in Fig. 23. During the early stages of redistribution, the water near the surface drains while the wetting of the dry soil beneath proceeds stepwise. However, the initial pattern tends to become smoother later.

If infiltration proceeds for a long time, the soil water profile may be connected with the capillary fringe above the water table of the ground water. Then water is removed from the profile as drainage water. In the absence of water removal by roots or evaporation, the final profile of soil water resembles that shown in Fig. 27a. A region of uniform water content exists above the water table where flow is caused solely by means of gravitational potential gradients.

In the absence of vegetation, evaporation occurs directly from the soil. When a water table exists close to the soil surface, the patterns of water redistribution are as shown in Fig. 27b. This pattern changes in the absence of a high water table. In this case, under constant external conditions, drying occurs in two stages (Fig. 27c): (1) an

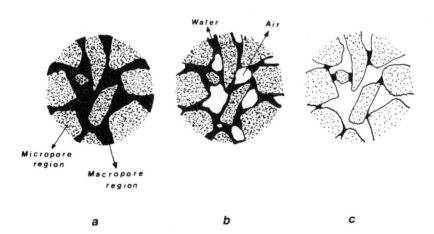

Fig. 26. Various stages in the dehydration of an aggregated soil. (a) Saturation. Both macro- and micropores are filled with water. (b) Air enters the soil and the continuity of water in the macropore region is broken, while micropores still remain saturated. (c) Further dehydration causes restriction of the macropore water to menisci between adjacent crumbs while micropores themselves start to desaturate (after Youngs 1965).

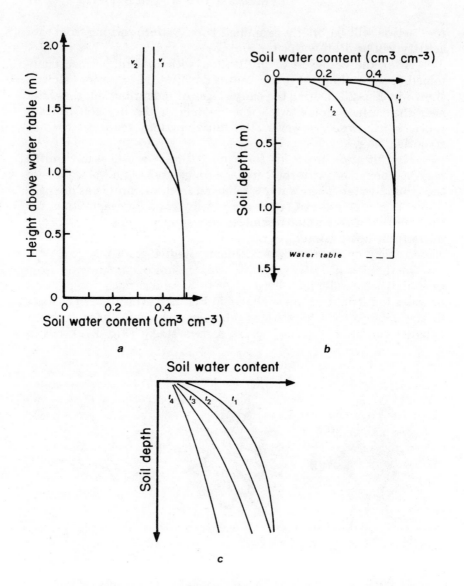

Fig. 27. Typical moisture profiles during redistribution of soil water. (a) During groundwater drainage in a slate dust for a water table at 2 m below surface. The two curves refer to two different final infiltration rates (v_1, v_2, where $v_1 > v_2$). (b) During evaporation in the presence of a water table at two different times t_1 and t_2. (c) During evaporation in the absence of water table (drying soil) with increasing time (adapted from Youngs 1965 and Hillel 1971). (Parts (a) and (b) reproduced by permission of the Society for Experimental Biology. Part (c) reproduced by permission from *Soil and Water* by D. Hillel. Academic Press Inc. (New York), 1971.)

early stage during which the evaporative demand rather than the soil properties determines the water profile, and (2) a second stage when actual evaporation from the soil is lower than the evaporative demand. Then, actual evaporation—and hence the water profile—is dictated by the ability of the soil to deliver water toward the soil surface.

Water-Vapor Movement Water movement in the vapor phase is insignificant at high soil water contents. However, as water is removed and void space increases, the proportion of water-vapor movement increases to become the dominant transport mechanism at a potential of about –15 bar (Slatyer 1967). In general the rate of vapor movement in soils (q_v) is thought to proceed by molecular diffusion along a vapor-concentration gradient (dc_w/dx). In a porous medium, the simple diffusion equation is modified as follows:

$$q_v = -\epsilon f D_w \frac{dc_w}{dx} \tag{29}$$

where ϵ is a tortuosity factor, f the fractional cross-sectional area available for diffusion, and D_w (cm^2 s^{-1}) the diffusion coefficient of water vapor into air. D_w is not constant but increases with temperature and decreases with the total pressure. The addition of the factor ϵf in the diffusion equation reduces the rate of diffusion in a porous medium such as soil by about 40% of that occuring in a free space. In practice, the values of q_v were found to be higher than those predicted from Eq. 29 (Gurr et al. 1952). This discrepancy was attributed to the presence of liquid water in the soil, which facilitates vapor flow (Philip and de Vries 1957).

Vapor-concentration gradients in soil arise from temperature, water potential, or solute potential gradients. Of these three, temperature gradients are the most important because they are easily established, particularly near the soil surface. Thus vapor flow under isothermal conditions can be regarded as negligible.

THE PLANT

Water is the major component of the plant body. It constitutes about 80 to 90% of the fresh weight of most herbaceous plant organs and over 50% of the fresh weight of woody parts. Water affects markedly, either directly or indirectly, most plant physiological processes (Hsiao 1973). Hence, with the exception of some kinds of seeds, dehydration of plant tissues below some critical level is accompanied by irreversible changes in structure and ultimately by plant death. The importance of water in living organisms results from its unique physical and chemical properties (Bernal 1965), which also determine its functions in plant physiology: water is a major constituent of the protoplasm; it acts as a solvent for many solid and gaseous substances, forming a continuous liquid phase throughout the plant; it takes part in many important physiological reactions; it maintains cell turgor, which exerts an impact on many physiological processes; and so forth.

The status of water in plant cells and tissues, the transport of water through plants, and, finally, the mode of development of plant water deficits will be treated in detail in the following sections.

The Status of Water in Plants

The thermodynamic approach adopted for the expression of soil water status is also applied to the study of plant water relations (Slatyer and Taylor 1960). Thus, the plant water potential (Ψ_{plant}) as defined in Eq. 17 is used to express the energy status of water in plant cells and tissues.

Ψ_{plant} consists of three component potentials:

$$\Psi_{plant} = \psi_p + \psi_s + \psi_m \tag{30}$$

where ψ_p, ψ_s, and ψ_m are the pressure, solute, and matric potentials, respectively. To understand the nature and contribution of the component potentials in plant cells, let us consider a mature parenchyma cell (Fig. 28).

This cell consists of three distinct phases: an elastic cell wall, the parietal cytoplasm with the nucleus and the organelles, and a central vacuole containing a dilute solution of sugars, ions, organic acids, etc. The vacuole occupies about 80 to 90% of the total volume of such a cell and is surrounded by the tonoplast, a semipermeable membrane. It is therefore reasonable to suggest that cell water exchanges are controlled by the vacuole and, furthermore, as a first approximation, to consider that a mature parenchyma cell behaves like an osmometer. In such a situation the contribution of matrix is neglected ($\psi_m = 0$) and Eq. 30 becomes:

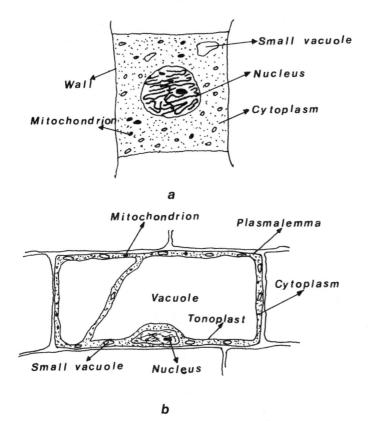

Fig. 28. Diagrammatic representation of (a) a meristematic and (b) a mature parenchyma cell from higher plants. The large vacuole of the mature cell controls its water exchanges in a manner close to that of an ideal osmometer. The same does not apply to the meristematic cell, where vacuolation is small or negligible. (With permission from *Plant-Water Relationships* by R. O. Slatyer. Copyright © 1967 Academic Press Inc. (London) Ltd.)

$$\Psi_{plant} = \psi_p + \psi_s \qquad (31)$$

The *solute potential* (ψ_s) is determined by the concentration of the osmotically active substances in the vacuole and is identical to the osmotic pressure of the vacuolar sap, but with a negative sign (Eq. 19). In a plant cell, ψ_s always bears negative values, which vary with cell volume. Thus ψ_s is closer to zero in fully hydrated cells than in dehydrated ones (because of the concomitant variations in cell volume), and the dependence of ψ_s on cell volume is approximately linear (Fig. 29).

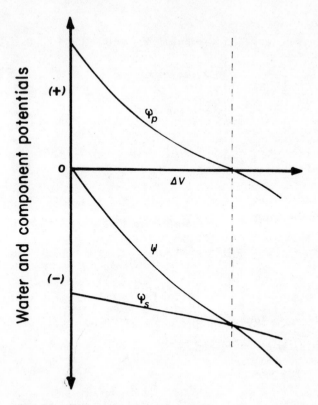

Fig. 29. The relationship between the volume of water lost (ΔV) and the water (ψ) and its component potentials, solute (ψ_s), and pressure potential (ψ_p), for a cell showing ideal osmotic behavior. The dashed line indicates the point of incipient plasmolysis (after Weatherley 1965b). (Reproduced by permission of the Society for Experimental Biology.)

The changes in cell volume are also responsible for the development of the *pressure potential* (ψ_p). When water enters the cell the vacuolar volume increases and a pressure, called turgor pressure, is exerted on the cell walls. At the same time a pressure equal to the turgor pressure is developed in the opposite direction, that is, from the walls to the cell interior. This latter pressure, called wall pressure, acts like a hydrostatic pressure, increases the energy status of water in the cell, and represents the cell pressure potential (ψ_p). Obviously, ψ_p takes positive values as long as the vacuole exerts a pressure on the surrounding walls. As water is lost from the cell, the vacuole contracts progressively, with a concomitant fall in cell turgor and ψ_p

in a curvilinear fashion (Fig. 29). ψ_p first becomes zero at a stage known as incipient plasmolysis, in which the vacuole ceases to press the walls (Fig. 29). Negative values of ψ_p arising from an inward pulling of the walls by the contracting vacuole have been occasionally reported (for example, Slatyer 1957), but these findings have been challenged by other investigators (Tyree 1976).

The pattern of the relationship between Ψ and the changes in cell volume is determined by the corresponding patterns of the component potentials (Höffler diagram, Fig. 29). When the cell achieves its maximum hydration (full turgor), Ψ bears its maximum value (zero) because $|\psi_p| = |\psi_s|$. As cell hydration falls Ψ drops to more negative values, and at incipient plasmolysis $\psi_p = 0$ and $\Psi = \psi_s$. The diagram of Fig. 29 leads to two ecologically important conclusions. (1) In view of the great significance of cell turgor to many physiological processes (Hsiao 1973), a maintenance of ψ_p above zero at relatively high levels of dehydration should be beneficial for plants growing in arid regions. This can be achieved by a more elastic cell wall, which makes the fall of ψ_p with increasing dehydration less abrupt. (2) It is possible that an accumulation of osmotically active substances takes place in the vacuole. This leads to a drop of ψ_s to values more negative than those expected by a simple volume reduction caused by cell dehydration. This solute accumulation in cells subjected to water stress constitutes an adaptive mechanism known as osmoregulation. The beneficial effects of osmoregulation are twofold. First, it enables cells to lose more water before their turgor drops to zero; second, it increases the ability of the cells to absorb water under dry conditions by lowering the cell water potential and thus maintaining a potential gradient between plant cells and their medium necessary for water movement toward cells.

The *matric potential* (ψ_m) arises in the presence of matrix and comprises forces retaining water molecules by capillarity, adsorption, and hydration (Dainty 1963). In the plant cell, matrix is mainly confined in the cell wall and the cytoplasm. In the former case, the interwoven cellulose microfibrils create numerous microchannels where water is held mainly by surface tension. In the cytoplasm, water is adsorbed on the various macromolecules and colloids. ψ_m was taken as negligible in the osmometer approach adopted for mature parenchyma cells. Such an assumption is reasonable in cases where cell matrix is a small fraction of the total cell volume as, for example, in thin-walled, non-aged cells (Gardner and Ehlig 1965; Kassam 1971; Karamanos 1978b). However, in tissues with a high proportion of matrix (for example, in xerophytes and in meristems of mesophytes),

ψ_m cannot be ignored and the osmometer approach no longer holds (Warren Wilson 1967c; Al-Saadi and Wiebe 1973). In any case, the effect of ψ_m becomes more pronounced as the water content of the tissue falls.

Water Transport through the Plant

Overall Aspects

In parallel to the direct evaporation of soil water, there is a continuous movement of water from the soil to the atmosphere through plants. This movement occurs from higher to lower water potentials and is regarded as obeying an equivalent of Ohm's law (Van den Honert 1948), according to which:

$$q = \frac{1}{A}\frac{dV}{dt} = -\frac{\Delta\Psi}{r} \tag{32}$$

where q is the rate of water flux (the volume of water, V, passing through a unit area, A, per unit time, t), $\Delta\Psi$ the potential difference, and r the resistance across the pathway. For a better understanding of the factors that control water transport, the whole pathway can be regarded as consisting of different segments. Then, in the steady-state, the flow of water through the various segments is given by the expression:

$$q = -\frac{\Psi_1 - \Psi_2}{r_1} = -\frac{\Psi_2 - \Psi_3}{r_2} = -\frac{\Psi_3 - \Psi_4}{r_3} = \cdots \tag{33}$$

For most of the pathway, that is, from the soil up to the leaf mesophyll cells, water moves in the liquid phase. Then evaporation takes place in the substomatal cavities, and vapor-phase movement occurs in the very last segment of the pathway. In the following sections, the whole pathway will be divided into segments to allow water movement to be treated in more detail.

Water Movement through the Soil-Root System

Water flux from the soil to the root xylem is expressed by the relation:

$$q = -\frac{\Psi_s - \Psi_r}{r_s + r_r} \qquad (34)$$

where Ψ_s, Ψ_r are the soil and root xylem water potentials and r_s, r_r are the resistances to water flow through the soil and root cortex, respectively. Obviously, the soil resistance r_s is the reciprocal of the soil hydraulic conductivity K, and hence it strongly depends on soil structure and texture as well as on soil water content (Fig. 25). The root resistance r_r arises in the pathway from root epidermis across the cortex to the xylem vessels (Fig. 30).

It is usually considered that water crosses the root cortex mainly, if not entirely, by moving in the free space, that is, through cell walls

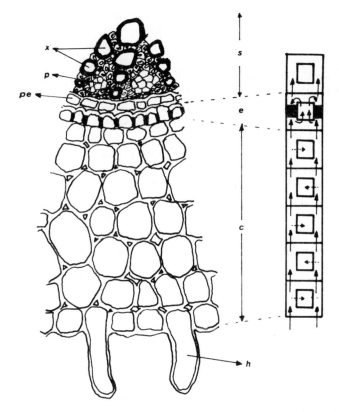

Fig. 30. Transverse section of a root in the zone of root hairs (h). c: cortex, e: endodermis, s: steele, pe: pericycle, p: phloem, x: xylem vessels. The diagram on the right indicates the pathway of radial water movement. Water is thought to move through the cell walls and intercellular spaces with minor exchanges with the symplast (dotted arrows). The Casparian strips of the endodermis induce a temporary entrance of water in the symplast of the endodermal cells.

and intercellular spaces (Weatherley 1963). The free-space movement stops at the endodermis because the radial cell walls are impermeable at the Casparian strip, and thus water enters the protoplast. It passes out again by the inner surface of the endodermal cells, and by further cell-wall movement it reaches the vessels. This pathway is regarded as offering the lowest resistance to water movement. However, other investigators (Newman 1976) consider that the symplasmic pathway, namely the water movement in the cell protoplasts via the plasmodesmata, may be of equal importance to the free-space movement. Root resistance was found not to be constant, but to decline with increasing water flux (Tinklin and Weatherley 1966), although the mechanism responsible for such a variability has not yet been located.

The relative magnitude of the resistances r_s and r_r is a matter of controversy. Recent experimental evidence emphasizes the greater significance of root resistance compared to soil resistance (Lawlor 1972; Reicosky and Ritchie 1976), in contrast with earlier studies (Gardner 1964; Cowan 1965) which attributed the greatest importance to the soil resistance. It appears, however, that the relative importance of these two resistances changes with soil water content. r_s is equivalent to or higher than r_r in very dry soils because of the shape of the relation between soil water content and hydraulic conductivity (Fig. 25). On the other hand, r_r is more important in nondry soils.

The water potential difference between root and soil ($\Psi_s - \Psi_r$) plays a critical role in the uptake of water by plants. For a continuous water supply, it is necessary that the root water potential (Ψ_r) is always more negative than the soil water potential (Ψ_s) (Eq. 34). Thus for an unimpaired supply of water to plants in a drying soil, the fall in Ψ_s must be accompanied by an equivalent fall in Ψ_r. However, there is a limit to the dehydration that plants can tolerate, and hence there is also a lower limit that Ψ_r cannot exceed. Both xerophytes and mesophytes have developed adaptive mechanisms (for example, osmoregulation) that enable them to decrease their water potentials to very negative values in order to sustain the potential gradient necessary for water absorption from very dry soils.

Equation 34 can be used as a basis to calculate water absorption by root systems following the approach of Gardner (1964). Accordingly, the rhizosphere can be considered as a number of layers where uptake follows the relation of Eq. 34. The soil resistance r_s is then taken as:

$$r_s = \frac{1}{BKL} \tag{35}$$

where K is the soil hydraulic conductivity, L a function of root distribution (that is, the length of root per unit soil volume), and B a geometrical factor independent of root distribution.

Gardner omits r_r from Eq. 34 because he considers it to be much smaller than r_s. Thus, for a given soil layer i, Eq. 34 becomes:

$$q_i = -BK_iL_i(\Psi_{s,i} - \Psi_{r,i}) \tag{36}$$

and for the whole root system:

$$\Sigma q_i = -B \sum_{i=1}^{n} K_iL_i(\Psi_{s,i} - \Psi_{r,i}) \tag{37}$$

Equation 37 predicts soil water uptake by root systems with considerable accuracy (Fig. 31) except for deep soil layers, where the

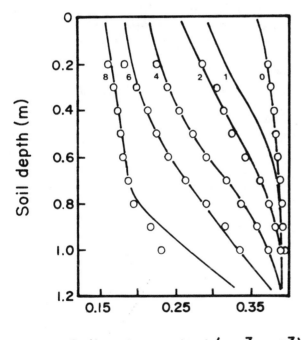

Fig. 31. Moisture profiles in the root zone of a sorghum crop. The numbers on the curves are days after irrigation. Circles represent experimental data, and the curves have been derived from Eq. 37 (after Gardner 1964).

elimination of r_r may not be absolutely justified in the light of more recent evidence (Willat and Taylor 1978; Karamanos 1980).

Water Movement through the Vascular System

Water movement from the root steele up to the leaf mesophyll cells occurs through the plant vascular system, which consists either of tracheids (gymnosperms) or mainly of xylem vessels (angiosperms). The driving force for the upward movement of water is the negative hydrostatic pressure (tension) that develops in the xylem of transpiring plants. Except for halophytes, the solute content of the xylem sap is usually very low (0.2 to 0.4%), and thus its solute potential is very close to zero. Water forms a continuous liquid system from the soil up to the evaporating surfaces in the leaf mesophyll. When water evaporates from the leaf cells, the reduction in potential at the evaporating surface (the leaf cell walls) causes movement of water from the xylem to the evaporating surface, which in turn reduces the pressure of water in the xylem and causes the ascent of water. This reduced pressure is transmitted throughout the liquid continuum to the root surfaces, where it reduces the root water potential and causes uptake of water from the soil.

To describe quantitatively water movement in the xylem vessels (the flow through tracheids is more complicated), the vessel is taken as a cylindrical microtube of radius R (usually between 10 and 500 μm, with an average of 20 μm) where water flow is assumed to be laminar. Then, according to Poiseuille's law,

$$q = -\frac{R^2}{8n}\frac{d\psi_p}{dx} \qquad (38)$$

where n is the viscosity of water (in poise or dyne s cm^{-2}) and $d\psi_p/dx$ is the driving force caused by the negative hydrostatic pressure. In practice, water flow in the xylem vessels was found to diverge from the values predicted by Poiseuille's law (Giordano et al. 1978).

The continuity of water columns in the xylem is necessary to ensure uninterrupted water supply to the transpiring organs. Dixon (1914) showed that water has high cohesive forces between 100 and 200 bar, especially when confined in small tubes with wettable walls. The role of the cohesive forces in maintaining the water columns unruptured can be appreciated when considering that the tensions developed inside the xylem do not exceed -200 bar even in xerophytes. However, breakage of water columns can easily occur either by dis-

TABLE 6: Estimated Relative Resistances of Liquid Water Movement through Roots, Stems, and Leaves of Sunflower and Tomato

	Resistances (s cm^{-1})	
	Sunflower	Tomato
Leaves	0.9	0.6
Stem	0.4	0.24
Roots	1.5	1.0
Whole plant	2.4	1.4

Source: Cited by Kramer (1969).

solved gases which form bubbles or by winds which produce cavitations. Both kinds of disturbances are usually confined in the larger vessels, while the smaller ones are usually filled with water. Nevertheless, blockage of the vessels by entrapped air bubbles is not easy, because of the presence of many cross-walls that retain the bubbles within the individual xylem elements and prevent the formation of larger bubbles and the complete blockage of the elements.

In comparison with the other tissues of roots and stems, xylem offers the lowest resistance to water movement (Table 6), and so practically all longitudinal water movement occurs in it. According to some scientists, the development of the vascular system constitutes the greatest single evolutionary step in the colonization of land because it allows a relatively unimpeded water supply to the overground plant parts. In the absence of a vascular system, the water supply to the transpiring organs would be more restricted since the resistance to water flow in the free space is several times greater than that in the vascular elements (Slatyer 1967; Kramer 1969). Accordingly, the risks for desiccation would be more pronounced even in low or moderate evaporative demands. In addition, little vertical extension of the plants would be permissible, in order to keep the overground plant part as close as possible to the water-supplying surface.

In parallel to the upward flux in the transpiration stream, there is also lateral water exchange between the xylem and the surrounding tissues, mainly occurring through the vessel pits. Such a lateral exchange can be demonstrated by comparing diurnal variations in stem diameter, with the tension in the xylem expressed as xylem water potential (Fig. 32). It can be seen that stem shrinkage occurs with increasing water stress, which indicates lateral water movement to the xylem vessels, while water moves in the opposite direction when xylem tension is decreased.

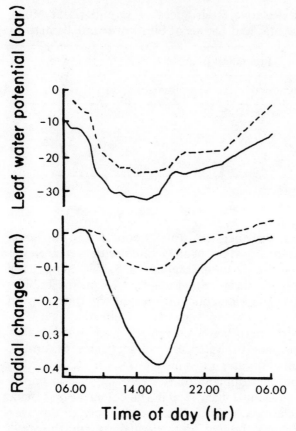

Fig. 32. Diurnal changes in stem radius and leaf water potential of cotton under wet (ψ_{soil} = -0.5 bar, dashed line) and dry (ψ_{soil} = -7.7 bar, continuous line) soil water conditions. ψ_{soil} was taken as the average for the 0 to 90 cm depth (after Namken et al. 1969).

Water Movement in the Leaves

Liquid water movement in the leaves occurs first in the vascular system through the veins. Water then moves through the mesophyll cells toward evaporating surfaces.

The arrangement of the vascular system varies widely in various kinds of leaves. Venation is usually characterized either as reticulate or as parallel. In the former case (dicots), veins form a branching pattern with successively thinner veins diverging as branches from the thicker ones. In parallel-veined leaves (monocots), strands of relatively uniform size are oriented in the blade longitudinally side-by-side, gradually converging at the apex. Nevertheless, the longitudinal veins are interconnected by smaller ones, so that even parallel-veined leaves show a reticulate arrangement at the microscopic level. The number of conducting elements decreases progressively with increasing distance from the main vein, and the ultimate branches consist of single xylem elements buried in the mesophyll. According to Kramer (1969), the actual distribution of water in the mesophyll occurs chiefly from the smaller veins, which are so numerous that most cells of a leaf are only a few cells away from a vein or vein ending.

Water movement outside the vascular system occurs in two directions: toward the epidermis and toward the mesophyll. Transport to the epidermis takes place either in the liquid phase, mainly through the bundle sheath extensions, or in the gas phase after the evaporation in the substomatal cavities. Transport in the mesophyll occurs in the liquid phase via the cell walls and intercellular spaces toward the evaporating surfaces in the substomatal cavities. As in the root cortex, the apoplastic pathway for water movement offers about 50 times lower resistance than the movement through the protoplasts (Weatherley 1963).

Water Movement in the Vapor Phase

Water movement in the vapor phase occurs in the very last segment of the soil–plant system and constitutes the physiological process known as transpiration. The radiation balance of a leaf at a given time determines the potentiality for the conversion of liquid water into vapor. Thus the radiative properties of a leaf (mainly its surface albedo) play a role in the evaporation rate of water in the mesophyll.

The evaporation sites are located both in the substomatal cavities and in the outer cell walls of the epidermal cells, unless strong secondary thickening is present. In the former case, evaporation mainly occurs in the inner epidermal wall in close proximity to the stomatal pore (Fig. 33) and not from the walls of the mesophyll cells which line the cavities. This is suggested from the larger vapor-density difference created at these sites (Meidner and Sheriff 1976) as well as from the internal suberization of the walls of mesophyll cells (Scott 1950).

The rate of vapor transport from the evaporating surfaces to the free air, the transpiration rate (T), is given by a relation similar to that used for the rate of evaporation (Eqs. 6 and 7):

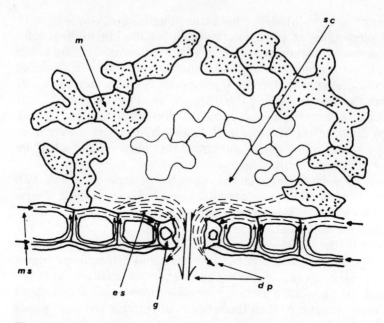

Fig. 33. Diagrammatic representation of vapor flow lines between leaf air spaces and the outside atmosphere according to Meidner and Sheriff (1976). *sc*: substomatal cavity, *m*: mesophyll cells, *es*: suggested main evaporation sites, *g*: guard cell, *dp*: outward diffusion pathway, *ms*: main stream of liquid water. (Reproduced by permission from *Water and Plants* by H. Meidner and D. Sheriff. Blackie and Sons Limited, 1976.)

$$T = -\frac{c_1 - c_a}{r_g + r_a} = -\frac{273\rho_v}{PT}\frac{e_1 - e_a}{r_g + r_a} \tag{39}$$

where the symbols are the same as in Eqs. 6 and 7.

In contrast with the liquid water flow in other segments of the continuum, the driving force of vapor transport is the vapor concentration rather than the potential difference between the leaf evaporating surfaces and the free air. However, Eq. 39 can also be expressed in terms of water potential difference by taking into account that the water potential of the air (Ψ_a) can be calculated from the relation:

$$\Psi_a = \frac{RT}{\overline{V}_w}\ln\frac{e}{e_s} \tag{40}$$

where e and e_s are the actual and saturation vapor pressures at a given temperature, respectively. To visualize the magnitudes of Ψ_a, let us consider a situation when air is saturated by 70% at a temperature of 20°C. At that temperature, e_s = 23.3 mbar, e = 16.3 mbar, and $RT/\overline{V}_w$ = 1,355 bar. Then, by substituting in Eq. 40, we find that Ψ_a = -484 bar. Obviously, Ψ_a becomes more negative with both increasing saturation deficit and increasing temperatures. The very negative Ψ_a in comparison with the values of water potential usually encountered in plant tissues indicates that the largest potential gradient in the whole system occurs in the leaf-to-air segment. This observation highlights the importance of a precise but flexible control of water-vapor flux in that segment of the pathway.

Leaf temperature determines the value of c_1 or e_1, given that the vapor concentration in the substomatal cavities is taken as that of saturated air. Alternatively, c_a and e_a depend on both air temperatures and dryness.

The resistances related to transpiration require special attention. The leaf gaseous resistance (r_g) refers to the parallel vapor flow through stomata and cuticle. Thus, r_g is thought to consist of two resistances connected in parallel (see also Fig. 37):

$$r_g = \frac{1}{r_{st}} + \frac{1}{r_c} = \frac{r_{st} + r_c}{r_{st} r_c} \tag{41}$$

where r_{st} is the stomatal resistance and r_c the cuticular resistance.

The *stomatal resistance* depends on the number of stomata per unit leaf area as well as on the geometry of the stomatal pore. Thus, for a given plant species, r_{st} is related to the degree of stomatal opening. The variations in stomatal aperture are achieved by turgor changes of the guard cells which line the stomatal pore. Because of their nonuniform wall thickness (Fig. 34), the guard cells are bent as their turgor (water content) increases, and this results in the opening of the pore.

The opposite occurs as guard cells lose water. The fluctuations in guard-cell water content which bring about the changes in their turgor are caused by means of an active potassium pump (Fischer 1968). Potassium cations are pumped into the guard cells from the surrounding subsidiary or epidermal cells when stomata are about to open. Such an influx of ions decreases the solute potential of the guard cells in comparison with that of the epidermal cells, thus causing a net water influx to the guard cells. On the other hand, the pump is inactivated when stomata are about to close, which causes a passive movement of K^+ out of the guard cells.

Stomata respond to a number of environmental factors such as light, relative humidity, CO_2 concentration, temperature, and leaf water status. (1) Stomatal resistance decreases with increasing light intensity, reaching a minimum value at a "saturating intensity" (I_m, Fig. 35a). (2) r_{st} increases with increasing ambient CO_2 concentration (Fig. 35b). (3) Increasing atmospheric dryness increases r_{st} (Schulze et al. 1972) (Fig. 35c). (4) r_{st} decreases with increasing temperature up to an optimum value beyond which stomata start to close (Fig. 35d). (5) Stomata close when leaves reach a critical level of dehydration (Fig. 35e). This enables plants to avoid desiccation damage by eliminating water loss.

Of the factors mentioned above, the concentration of CO_2 is thought to exert the most important influence on stomatal movements, while light is not essential for stomatal opening (Raschke 1976), provided that leaf water potential is above a threshold value. If dehydration continues and the threshold value is reached, then stomata close irrespective of the CO_2 concentration. This switch-like response is thought to be caused by the accumulation of abscisic acid in the guard cells, a growth inhibitor that causes immediate stomatal closure (Hiron and Wright 1973). In this way, stomata solve a dilemma of opposing priorities: to remain open for a more rigorous assimilation of CO_2 and, at the same time, to prevent excessive water loss through transpiration.

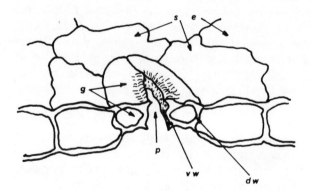

Fig. 34. Section of a kidney-shaped stoma shown as a perspective diagram (g: guard cell, p: pore, s: subsidiary cells, e: epidermal cell, vw: thick ventral wall, and dw: thin dorsal wall of the guard cell) (after Meidner and Mansfield 1968.) (From *Physiology of Stomata* by H. Meidner and T. A. Mansfield. McGraw-Hill (UK) Ltd., 1968. Reproduced by permission of the authors.)

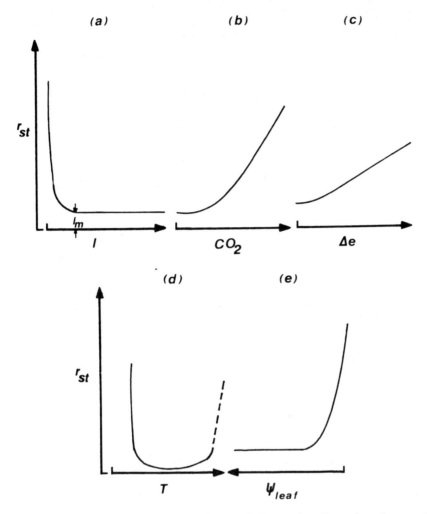

Fig. 35. Changes in stomatal resistance (r_{st}) as a function of environmental and plant factors. (a) Light intensity (I). I_m represents the saturating intensity. (b) CO_2 concentration. (c) Vapor concentration difference between mesophyll and the atmosphere (Δe). (d) Temperature (T). (e) Leaf water potential (ψ_{leaf}).

There are enormous differences among plants with respect to the response of their stomates to environmental factors. Some factors such as the threshold value of leaf water potential at which stomata close are of great ecological importance, since this value determines the degree of dehydration that plants can withstand before they stop water loss. A general idea about the potential of each species to transpire can be obtained by comparing the lowest values of r_{st}, that is, the values above the saturating light intensity in well-watered plants. It can be seen from Table 7 that the minimum values of r_{st} differ even within mesophytes.

The *cuticular resistance* depends on characteristics of the leaf cuticle. A high r_c is associated either with thicker cuticles or with cuticles enriched with hydrophobic materials. r_c is not flexibly controlled by the plant as is r_{st}. Nevertheless, it was found to decrease with increasing temperature (Holmgren et al. 1965) and increasing atmospheric relative humidity (Moreshet 1970). When compared with the minimum values of r_{st}, r_c is higher by about 1 to 2 orders of magnitude (Table 7). Thus, cuticular transpiration is negligible in comparison with stomatal transpiration during the day. The ecological importance of r_c is more apparent at night when stomata close: xerophytes achieve much higher values of r_c than those of mesophytes (Table 7), so that they have lower rates of cuticular transpiration.

The resistance r_a, called *boundary layer resistance*, has the same origin as the aerodynamic resistance of natural surfaces referred to earlier. r_a is proportional to the thickness of the sheath of still air (d) in contact with the leaf surface:

$$r_a = \frac{d}{D_w} \tag{42}$$

where D_w is the diffusion coefficient of water vapor to the air (about 0.24 cm^2 s^{-1}). Windspeed strongly affects the thickness of the boundary layer: d is highest in still air and decreases with increasing

TABLE 7: Minimum Values of Stomatal Resistance (r_{st}) for Various Mesophytes

Species	r_{st} (s cm^{-1})
Turnip, sugar beet	1.5 to 1.7
Bean, tomato	2.3 to 3.3
Xanthium	0.8 to 1.7
Cotton	0.9 to 1.3
Barley	1.0 to 2.0

Source: Cited by Monteith (1965).

windspeed. Monteith (1965) suggested the following expression for the calculation of r_a:

$$r_a = 1.3 \sqrt{\frac{l}{u}} \tag{43}$$

where l is the leaf width and u the windspeed. Usually r_a varies between 0.1 to 3 s cm^{-1} for a normal range of windspeeds and leaf shapes (Slatyer 1967). The magnitude of r_a in still air (about 2 to 3 s cm^{-1}) in comparison with the minimum values of r_{st} usually encountered suggests that r_a may play a significant role in controlling transpiration under these circumstances (Fig. 36). Some morphological characteristics of leaves such as hairiness and folding are regarded as adaptive responses associated with an increased r_a.

Conclusions

When considering water flow throughout the soil-plant-atmosphere continuum, according to van den Honert (1948), the flow equations for the separate segments (Eq. 33) are converted as follows:

$$q = -\frac{\Psi_s - \Psi_r}{r_s + r_r} = -\frac{\Psi_r - \Psi_x}{r_x} = -\frac{\Psi_x - \Psi_1}{r_1}$$

$$= -\frac{\Psi_1 - \Psi_a}{r_g} = -\frac{\Psi_s - \Psi_a}{\Sigma r} \tag{44}$$

Thus the flow of water in the system can be treated as analogous to the flow of an electric current through a series of resistances and capacitances (Fig. 37). The capacitances represent the ability of the various segments of the pathway to dehydrate and rehydrate in response to the tissue water balance. A tissue behaves as water storage which can supply the main pathway with water during the stress periods. Since this water supply lags behind the fluctuations in transpiration, it is represented as occurring through lateral resistances connected with the capacitance. The water-storage capacity of the plant tissues in a form analogous to hydraulic capacitance can be understood when considering the fluctuations in the dimensions of different plant organs related to their water status. Thus shrinkage and swelling in response to the internal water balance has been observed in leaves (Gardner and Ehlig 1965), stems (Namken et al. 1969), roots (Huck et al. 1970), and fruits (Chaney and Kozlowski 1971).

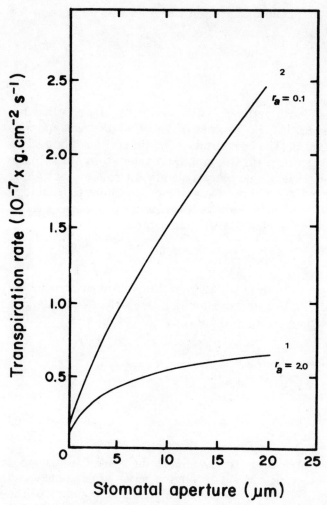

Fig. 36. The stomatal control of transpiration of *Zebrina* leaves in quiet air (curve 1, $r_a = 2.0$ s cm^{-1}) and in moving air (curve 2, $r_a = 0.1$ s cm^{-1}) (after Bange 1953). Stomata control transpiration effectively when the r_a is small; the same does not apply in quiet air, where r_a is often as large as the stomatal resistance.

As previously stated, the existence of a very steep potential gradient in the leaf-to-air portion of the pathway requires a decisive control system of the flow to prevent tissue dehydration. From that point of view, stomata are strategically placed (Weatherley 1976) to control vapor flow. Table 8 shows the magnitudes of the resistances to vapor flow in different plant species.

Roots offer the major resistance to the flow of liquid water, probably because of the existence of the endodermis (Table 6). On

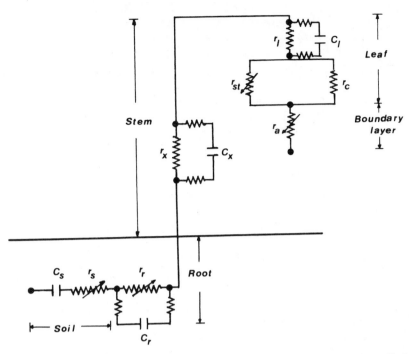

Fig. 37. Diagrammatic representation of the water movement in the soil-plant-atmosphere continuum in the form of a network consisting of a number of resistances (r) and capacitances (C). The symbols are the same as those used in the text.

TABLE 8: Resistances to Transfer of Water Vapor through the Boundary Layer (r_a), the Cuticle (r_c), and the Stomata (r_{st}) in Leaves of Several Species

Species	Resistances (s cm^{-1})		
	r_a	r_{st}	r_c
Betula verrucosa	0.80	0.92	83
Quercus robur	0.69	6.70	380
Acer platanoides	0.69	4.70	85
Circaea lutetiana	0.61	16.10	90
Lamium galeobdolon	0.73	10.60	37
Helianthus annuus	0.55	0.38	—

Source: Adapted from Holmgren et al. (1965).

the other hand, the stem offers the lowest resistance, and leaf resistance is intermediate between the two extremes.

Many of the resistances of the pathway can be characterized as adjustable in the sense that they undergo major variations in response to several factors. Thus soil resistance (r_s) is a function of soil water content, while the boundary layer resistance (r_a) is a function of windspeed. From the plant resistances, root resistance (r_r) varies with the transpiration rate, while stomatal resistance varies enormously in response to external and internal factors.

The Development of Internal Water Deficits

Cells and tissues are regarded as water-deficient when they are not at full turgor (Crafts 1968; Kramer 1969), that is, when their water potential is below zero. The term water stress is also used to express such a situation, and water potential is regarded as a suitable measure of stress since it has the same physical dimensions with mechanical stress, namely $ML^{-1} T^{-2}$ (Lockhart 1965). Since plant tissues never reach zero water potential even when grown in water culture (for a review, see Karamanos 1980), the definition of water stress given above is not realistic. In physics, the concept of stress (applied force) is bound with the concept of strain, that is, the deformation suffered in a physical system subjected to stress. Equivalently, in plants water stress must be related to drought-induced departures of physiological processes from the normal.

To understand the mode of development of plant water deficits, let us consider a simple physical model proposed by Dixon in 1938 (Weatherley 1976) (Fig. 38). In this model the plant is regarded as consisting of two porous pots filled with water and connected by a tube through a manometer. The lower pot is semipermeable, buried in the soil, and represents the root. The upper pot represents the leaf; the manometer represents the vacuoles of the mesophyll cells which are off the main pathway, since water moves mainly through the cell walls. At a steady rate of transpiration, the manometer will register a drop in pressure in the pathway. The existence of root resistance is reponsible for this pressure drop. Because of root resistance, water absorption from the lower pot cannot meet the transpirational fluctuations instantaneously; thus water will be added to or subtracted from the manometer, which will register the tension variation in the system. Accordingly, the tissue water potential is equivalent to the negative pressure registered by the manometer.

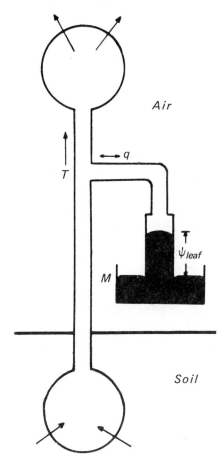

Air

q

T

ψ_{leaf}

M

Soil

Fig. 38. Dixon's model of a transpiring plant. The upper and lower porous pots represent the leaves and roots respectively, while the manometer (M) represents the mesophyll cells. The negative pressure registered by the manometer is equivalent to the leaf water potential (ψ_{leaf}) which fluctuates in response to lateral water movement (q) from and to the cells. T denotes the main transpiration stream (from Weatherley 1976).

Thus the water balance of a tissue at a given time results from the balance between absorption and transpiration. As soil water is removed and soil water potential is falling to more negative values, plants must decrease their water potential (either by simple dehydration or by solute accumulation) to sustain the potential gradient necessary for water absorption. At the same time, absorption becomes more difficult because of the great increase in soil resistance. Plant water potential thus becomes progressively more negative with increasing soil dryness (Fig. 39).

At night, when transpiration ceases, the daily water deficits are gradually eliminated and plant water potential eventually attains a

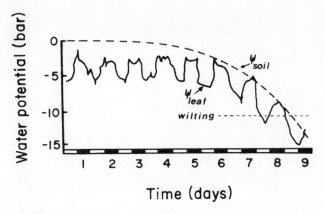

Fig. 39. Changes in leaf water potential (ψ_{leaf}) and soil water potential (ψ_{soil}) of a pepper plant grown in a clay loam soil under controlled evaporative conditions. The dotted line indicates the level of water potential below which leaf wilting occurs (after Gardner and Nieman 1964).

sort of equilibrium with soil water potential. Plant water potential undergoes a more-or-less typical pattern during the day, with its highest values at dawn and sunset and its minimum value at about midday when the imbalance between transpiration and absorption is at a maximum (Fig. 40). The amplitude of this daily trend increases with increasing soil dryness. When soil water potential falls below a critical value which represents the ultimate ability of the plant tissues to withstand a further decrease in their water potential (dashed line in Fig. 39), absorption ceases and permanent wilting occurs.

Results from several experiments (Makkink and van Heemst 1956; Denmead and Shaw 1962) suggest that under field conditions the evaporative demand rather than the soil water status determines plant water status (Fig. 41a). Alternatively, the extent of the root system, which determines the capacity for water uptake, may also play an important role (Fig. 41b).

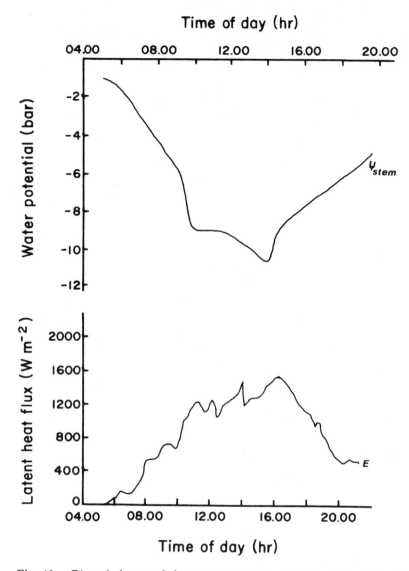

Fig. 40. Diurnal change of the stem water potential of nonirrigated apple trees (upper curve) and the latent-heat flux (calculated after Monteith) on a clear summer day at Long Ashton, Bristol (adapted from Powell 1974).

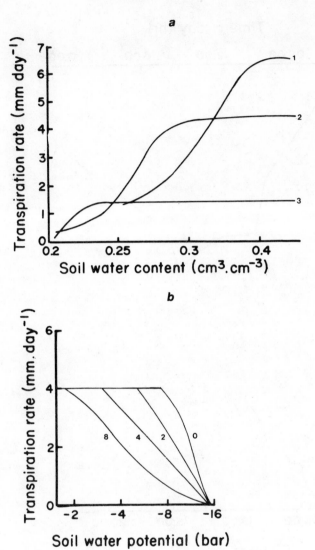

Fig. 41. (a) Effect of decreasing soil water content on transpiration rate on days of high (curve 1), moderate (curve 2), and low (curve 3) potential transpiration (after Denmead and Shaw 1962). Stomatal control of transpiration can easily occur even at high levels of soil water content when the potential evaporation is high. Alternatively, plants do not control transpiration under humid conditions even at low soil water content. (b) Expected effect of root density on the relationship between transpiration rate and soil water potential (after Cowan 1965). Root densities are 8, 4, and 2 cm^3 of soil per centimeter of root length, while 0 cm^3 is used for very dense rooting. Soil water potential is less limiting at high root densities than at low densities.

3

J. M. CLARKE
R. C. DURLEY

The Responses of Plants to Drought Stress

AVOIDANCE

The ability of a plant to grow satisfactorily when exposed to periods of water stress is called drought resistance (May and Milthorpe 1962). Drought resistance can take the forms of either the avoidance or the tolerance of severe levels of stress. Drought, or water stress, are relative terms and depend upon whether one is referring to plants classified as xerophytes, mesophytes, or hydrophytes. The present discussion will be limited largely to xerophytes and mesophytes.

A drought-avoiding plant, by definition, is one that maintains high plant water potential when exposed to extended water stress. For the purposes of this discussion a somewhat broader definition of avoidance will be used. Plants that escape or evade drought through the adjustment of their life cycles will be included as well. Such plants are commonly called drought evaders or drought escapers because they "avoid" drought by completing their life cycles before drought becomes too severe (in the case of annuals) or become dormant during droughts (in the case of perennials). The "true" drought-avoiding plants possess mechanisms to maintain favorable water status, either by conserving water or by their ability to supply water to above-ground organs even during drought stress.

The sections on Avoidance, Tolerance, and Growth were contributed by J. M. Clarke; the section on Hormonal Changes was contributed by R. C. Durley.

The avoidance of drought by adjustment of the growth period of the plant is common in regions having Mediterranean and subtropical climates where definite wet and dry cycles occur. In annual species, germination, growth, and most of the reproductive development occur during the more moist periods; the dry period is avoided by production of a dormant seed. In perennials, growth and reproductive development occur during conditions of favorable moisture, while dry cycles are passed in a dormant or near-dormant state.

In cultivated annual crops, the date of seeding is adjusted to make best use of available moisture. Additionally, cultivars with appropriate maturity have to be used so that the economically useful portion of the plant completes development prior to the onset of drought. In the case of wheat, for example, seeding takes place in the autumn in Mediterranean-type climates (dry summers, moist winters) so that the crop matures prior to, or at the start of, the drought (Klages 1942; Patterson 1965). Areas in which this is practical, in addition to the Mediterranean area itself, include the northwestern United States and Australia. In subtropical regions with bimodal precipitation distributions, seeding dates of wheat are determined by moisture availability and the need to avoid seed formation during excessively hot and dry periods of the year. These same principles apply to numerous other annual crops grown in hot regions experiencing periodic drought.

Wild plant species have successfully evolved life cycles and growth habits suited to their particular environments. The ephemeral plants found in desert regions provide a dramatic example of adaptation of the life cycle to available moisture. These plants germinate, grow rapidly, flower, and mature before the limited soil moisture is depleted. Ephemerals may be annuals reproduced by means of seeds, or perennials reproduced by bulbs or rhizomes. The seeds of annual ephemerals usually require at least 15 to 25 mm of rain before they will germinate. Mature plants may be only 1 to 2 cm tall; their size and length of growth cycle depends upon how long the soil moisture lasts.

Certain perennial species are also capable of avoiding drought, principally through a deciduous growth habit. In this regard it is important to distinguish between those species that lose their leaves as a part of their annual growth cycle and those that lose all or part of their leaf area in response to drought. Species in the latter category will be dealt with later in the discussion of mechanisms of water conservation. "Obligate" deciduous trees and shrubs in the Mediterranean region, for example, may be leafless during the dry summer and bear leaves and flowers during the rainy winter season (Oppenheimer

TABLE 9: Mechanisms of Water Conservation in Higher Plants

Mechanism	Mode of Action
A. Improved water uptake	extensive root systems
	efficient root systems
B. Control of transpirational loss	reduced leaf area
	cuticle changes to reduce water loss
	changes in leaf configuration and hairiness
	reduction of stomatal number and size; sunken stomata
	control of stomatal aperture
C. Water storage in plant tissue	stems, leaves, roots

1960). Such shrubs may have superficial root systems and do not depend upon deeply stored soil moisture to maintain active growth throughout the year.

A second means by which plants avoid drought is the conservation of available water to tide them over dry periods. This is accomplished by a wide variety of morphological adaptations and, in some cases, physiological adaptations as well (Table 9). The species that exhibit these mechanisms are considered to be true xerophytes, since they continue active growth during periods of severe drought. The means by which these plants conserve water can be divided into three basic areas:

1. improved water uptake from soil because of more extensive root systems,
2. control of transpirational losses of water from above-ground plants,
3. storage of water in plant tissues.

Extensive root systems to extract water from the soil are common in plants growing in drought-prone sites. This can take the form of deeper rooting if subsoil water is available, or extension of root systems near the surface where groundwater is not available (Fig. 42). Roots of the prairie rose (*Rosa arkansana*) growing in the Great Plains area of the United States may penetrate as deep as 4.5 to 6.5 m, while those of the Comanche cactus (*Opuntia camanchica*)

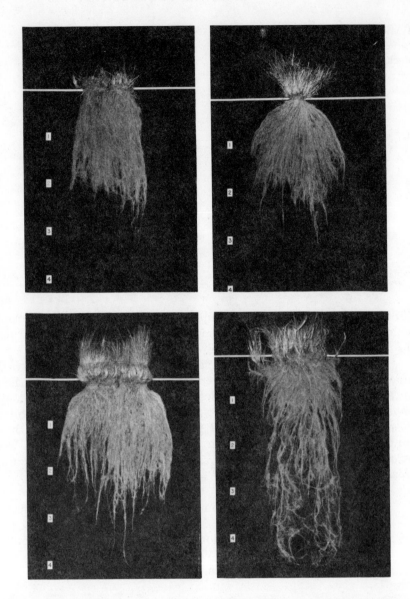

Fig. 42. Root systems of cultivars of forage grasses originating in a semiarid habitat (numbers are depths in feet). (a) *Agropyron intermedium* (intermediate wheat grass). (b) *Agropyron elongatum* (tall wheat grass). (c) *Elymus junceus* (Russian wild ryegrass). (d) *Elymus angustus* Trin. (altai wild ryegrass).

may penetrate only 3 to 8 cm but spread up to 2 m from the plant (Weaver 1926). The cactus is thus adapted to rapid absorption of water near the soil surface after rain. The morphological differences among plant species inhabiting a particular site have further ecological significance: they reduce root competition for space and water. Competition is reduced by having mixtures of shallow- and deep-rooted plants in the community.

The size of plant root systems is dependent upon both genetic and environmental factors. Studies of the root systems of range plants in Idaho revealed that annual species, such as *Bromus tectorum*, had few roots, and these penetrated only about 30 cm (Spence 1937). On the other hand, short-lived perennials, such as *Poa secunda,* had more abundant but still relatively short roots (about 40 cm). Longer-lived perennials, such as *Agropyron inerme*, had many more major roots, which penetrated to a depth of about 160 cm. Root systems of alfalfa vary from tap roots to rhizomes (Fig. 43). The extent of root development within species is governed by environmental factors such as moisture, soil structure, and light. Studies with crop plants generally indicate greater root growth in irrigated or moist environments than in dry environments (Mayaki et al. 1976), although this generalization does not hold in extremely wet environments. The ratio of root to above-ground dry matter, however, is generally higher for plants growing in dry habitats than for plants growing in moist habitats.

Investigations of the drought resistance of selections of smooth bromegrass (*Bromus inermis*) have demonstrated a positive relationship between drought resistance and depth of penetration and numbers of roots (Cook 1943). Deeper-rooting varieties of wheat produced higher yields under drought conditions (Hurd 1964) than shallow-rooting varieties. In a grass species (*Phalaris tuberosa*) of the Mediterranean region, deep roots are able to provide sufficient water to maintain the dormant plant during the summer drought period (McWilliam and Kramer 1968). In both native vegetation and cultivated crops, competition affects the spatial distribution and size of root systems. The relative size and depth of roots of plants in natural plant communities affects the ability of different species to survive. Some plants (for example, guayule) are able to enforce greater plant spacing by means of toxic root exudates, which prevent seedlings from developing too close to established plants.

Another series of mechanisms whereby plants avoid the effects of drought stress relates to control of transpirational water loss from above-ground plant parts. This is accomplished by changes in the surface area of transpiring plant parts, physical changes in transpiring

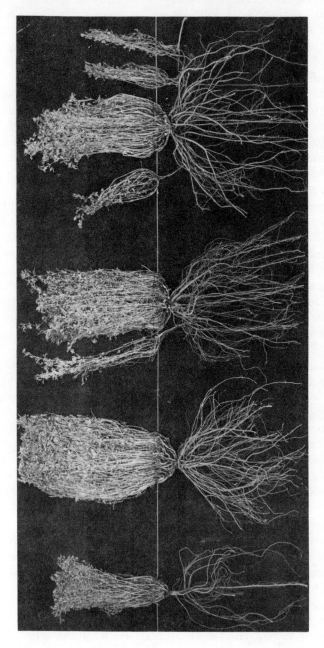

Fig. 43. Genetic variation in root systems of alfalfa (*Medicago* spp.), from tap-root system (left) to rhizomatous (right).

surfaces to limit water loss, and control of the opening of stomata. Plants growing in xeric habitats exhibit varying combinations of these mechanisms.

The reduction of the transpiring surface area is a common means of reducing water loss. This can take the form of reduced leaf area, either genetic or drought-induced, or changed leaf shape, such as leaf rolling in response to water deficits. Smaller leaves can have a more favorable energy budget, producing a lower transpiration rate (Gates et al. 1968).

Certain plant species will either accelerate natural leaf drop or undergo leaf drop unrelated to normal growth habit when exposed to water stress. The natural leaf loss in wheat plants, for example, is accelerated under water-stress conditions. The shrub *Salvia triloba*, which grows in the Mediterranean region, has large leaves in the rainy winter season but replaces them with smaller leaves during the dry summer (Oppenheimer 1960). In still other species, leaf rolling helps to reduce transpiration. This response is seen in many grasses. In the grasses *Agropyron dasystachyum* and *A. smithii*, leaf rolling increases as water stress develops, and rolling is most pronounced during midday when stress is greatest (Redmann 1973; Ripley and Redmann 1976). Leaf rolling serves to reduce both leaf area and radiant heat load. Reductions in transpiration rate of 46 to 83% due to leaf rolling have been found in some Mediterranean grasses (Oppenheimer 1960).

Some xerophytes have no leaves; the function of the leaves is taken over by twigs or stems. In the cactus family, most species have no leaves at all; in others, leaves exist only in the early stages of growth, and then only as small scales at the stem nodes. *Laurena arborescens*, a spiny bush of the *Compositae* family, has no leaves on its outer branches, whereas its inner branches, which are shaded by the outer branches, do have leaves (Grenot 1974).

Leaf or stem pubescence (hairiness) is often cited as a feature of plants adapted to arid environments. Pubescence can reduce the radiant heat load of leaves by increasing the reflectance of the leaf surface. The pubescence may also reduce the absorption of photo-synthetically active radiation, as demonstrated in a shrub genus in California (Ehleringer et al. 1976). Absorption was reduced by as much as 56% in *Encelia farinosa* (pubescent) as compared to *E. californica* (nonpubescent). The photosynthetic rate was lower in *E. farinosa* as well. There is conflicting evidence on the effect of pubescence on transpiration rate.

Considerations of morphological features of xerophytic plants usually also include the thickness of the leaf cuticle (outer layer) and the presence of surface coatings. Cuticular thickness alone is probably not an indicator of reduced leaf-water loss, since there are variations in the structure and chemical composition of the cuticle (Parker 1968). In some plant species, water loss through the cuticle can make up a significant portion of total leaf transpiration. In mesophytic oaks, cuticular transpiration has been shown to comprise 1/2 of total transpiration, while in laurel, which has more xerophytic characteristics, it comprises only about 1/45 of the total (Oppenheimer 1960). Depositions of lipids and other fatty materials in the cuticle help to reduce water loss. Waxes and resins excreted to the outer surfaces of leaves in some species may also aid water retention.

The stomata that occur in leaves, and in the stems of some species of plants, are the major pathway for gaseous exchange between the plant and the atmosphere. Carbon dioxide is taken in through stomata for use in photosynthesis, while water vapor passes out into the atmosphere. There is a duality in the nature of stomatal action: stomates must be open to permit photosynthesis and evaporative cooling of the leaf, but closure permits the conservation of water under drought conditions. Plant species that close stomata to conserve water must deal with higher internal temperatures, which may rise above ambient air temperature when the radiation load is high. Other adaptive features include reductions in numbers of stomata per unit surface area, and "sunken stomata." Sunken stomata, located in subepidermal cavities, increase the resistance of water flow from leaf to atmosphere by increasing the boundary layer, or "dead-air" space, around the leaf surface.

The rate of diffusion of CO_2 into the leaf, or of water vapor out of the leaf, is dependent on the size and distribution of stomata. In addition, water-vapor loss is affected by atmospheric humidity, radiation, and wind. General observations indicate that stomatal closure occurs during the early stages of water deficit, perhaps before visible symptoms such as leaf wilting occur (Meidner and Mansfield 1968). In some species, the reopening of stomata may be delayed for a short time after the water stress is relieved. Stomata of maize plants exposed to a severe drought lasting 7 days or more did not open as fully when rewatered, despite the apparent recovery of leaf turgidity (Glover 1959). The stomata of sorghum, on the other hand, reopened completely. Some drought-resistant plants keep their stomata open longer under stress conditions than some less drought-resistant plants, which allows photosynthetic production to continue longer.

Stomatal closure as a water-conserving mechanism usually induces a lower rate of photosynthesis. The perennial grass *Agropyron smithii*, when growing under drought stress, opens its stomata during periods of low radiation and temperature and high humidity just after sunrise (Maxwell and Redmann 1974). Photosynthesis can take place until stress levels increase to the point where stomata close. They remain closed throughout the remainder of the day and night. Similar responses are evident in sesame (*Sesamum indicum*) (Hall and Kaufmann 1975) and rape (*Brassica napus*) (Fig. 44). Other plants, such as sunflower (*Helianthus annuus*) give priority to maintaining photosynthesis, relying on an efficient water-transport system to maintain plant water status at acceptable levels (Hall and Kaufmann 1975). A perennial shrub, *Tidestromia oblongifolia*, growing in Death Valley, California, was found to carry on its highest rate of photosynthesis during the hot midday period (Bjorkman et al. 1972). This ability to photosynthesize under high stress levels was attributed to the use of the C_4 photosynthesis pathway by the shrub, which can continue at lower leaf CO_2 concentrations than with the C_3 pathway. The differences between the major photosynthetic pathways can be seen in Fig. 45.

The use of artificial antitranspirants has been suggested (see Chap. 4) as a way to help conserve water in crop plants. These substances may be agents that induce stomatal closure, film-forming compounds such as plastic films or waxes that cover the leaf surface, or reflective pigments that reduce radiation load and leaf temperature. The ideal antitranspirant would be one that permitted diffusion of CO_2 into the leaf but restricted the rate of water loss. In practice, most antitranspirants cause reductions in both photosynthetic and transpiration rates (Slatyer and Bierhuizen 1964). Reduced transpirational rates may cause an increase in leaf temperature to unacceptable levels.

The problem of closed stomata and their effect on gas exchange has been solved in a different way by plants that possess crassulacean acid metabolism (CAM). This metabolism allows chloroplast-containing cells to assimilate CO_2 in the dark, to synthesize malic acid which is stored in the cell, and subsequently to decarboxylate the malic acid in the following light period, which releases the CO_2 for refixation by photosynthesis (Osmond 1978). Thus the stomata of plants with the CAM pathway open at night to take in CO_2 when transpiration rates are low, and the leaves incorporate the fixed CO_2 into photosynthate the following day while the stomata remain closed. The pathway is found in numerous plant families and genera, including the cacti and one cultivated crop, pineapple. Growth rates are generally low in CAM plants, but growth rates of pineapple under favorable growth conditions may approach the lower range of growth rates found in C_3 plants (Osmond 1978). Some species normally exhibit C_3 photosynthesis but switch to the CAM pathway under

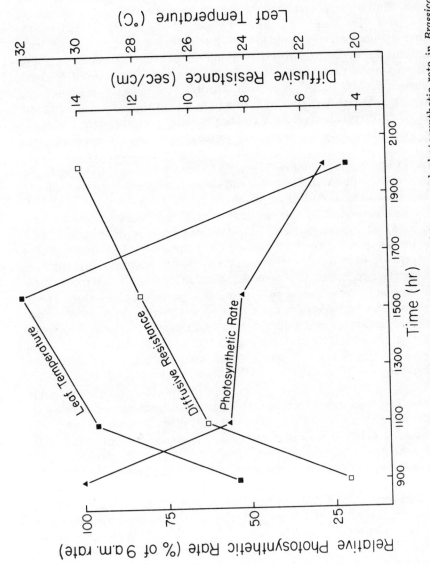

Fig. 44. Daily change in leaf temperature, diffusive resistance, and photosynthetic rate in *Brassica napus.*

stress conditions. Plants that normally use the CAM pathway may switch to the C_3 pathway and open their stomates during the day when growing in more mesic habitats. This latter behavior is found in *Agave deserti*, a species common in the Colorado, Mojave, and Sonoran deserts (Hartsock and Nobel 1976).

Although most of the major crop species tend to be mesophytes, some possess drought-avoidance characteristics. In sorghum, for example, differences in degree of leaf water stress have been found

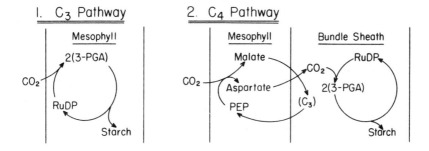

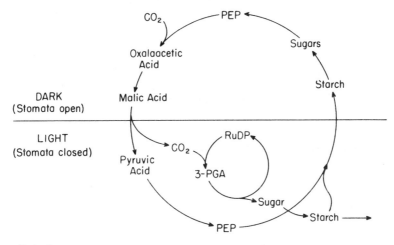

Footnote

PGA = Phosphoglyceric acid
PEP = Phosphoenol pyruvate
RuDP = Ribulose diphosphate

Fig. 45. Three alternative pathways for photosynthesis in higher plants.

between different varieties (Blum 1974). The major objective of plant breeding programs in crops for drought areas is to increase crop yield per unit of water used. This concept is termed water-use efficiency (WUE) and can be expressed as

$$\frac{\text{dry matter produced}}{\text{evapotranspiration of crop area}}$$

Improvements in WUE can be effected by optimizing management factors such as planting date and rate, and fertilizer application, as well as by breeding. The WUE of a crop reflects the integrated effects of root-system size and efficiency, water transport in the plant, and net photosynthetic output in relation to amount of water transpired. Water-use efficiency is greater in C_4 than in C_3 plants because of the higher photosynthetic rates and higher stomatal resistance of the C_4 plants. Another measure of the efficiency of water use is the transpiration ratio, which is the inverse of WUE, that is,

$$\frac{\text{g water transpired}}{\text{g dry matter produced}}$$

The transpiration ratio is generally lowest in CAM plants, intermediate in C_4 plants, and highest in C_3 plants (Table 10).

TABLE 10: Transpiration Ratios of Plants Using the C_3, C_4, and CAM Photosynthetic Pathways (g water transpired/g dry matter produced)

Photosynthetic Pathway	Species	Transpiration Ratio
C_3	wheat	435
	sunflower	569
	bean	538
	alfalfa	664
C_4	maize	230 to 270
	sorghum	270 to 380
	millet	260 to 370
CAM	pineapple	40 to 116
	agave	47 to 70
	Opuntia	50 to 150

Sources: Maximov (1929) and Ting (1976).

Another mechanism for drought avoidance is water storage in plant tissues to support the plant between rains. Plants with this capability are called succulents and may be divided into stem succulents, such as cacti, leaf succulents, such as aloe, and root succulents, such as *Parhypodium succulentum*. The succulents typically have rather shallow root systems, which are adapted to rapid absorption of water during and after rains. Another feature is slow growth rate, since photosynthesis is inhibited by the water-conservation measures. For this reason, succulents are generally poor competitors and are restricted to arid habitats, particularly habitats with two rainy periods per year (Walter and Stadlemann 1974).

In succulents, water is stored in enlarged parenchymous cells, which lose water during drought periods and are recharged during rainy periods (Walter and Stadlemann 1974). Although there may be a large loss of water during drought, the relative water content of the plant as a whole tends to remain fairly constant as a result of concomitant reductions in dry weight due to respiration.

The greatest proportion of the plant weight of succulents is comprised of the plant part that stores water, that is, the stem, leaf, and root of the above classes of succulents, respectively. In some root succulents, the weight of the underground tuber may be 10 times that of the shoot. Some succulents, such as the so-called soil cacti, exist almost entirely below the soil surface, with only the flattened leaf tips at the soil surface.

Drought avoidance by any particular plant species can be comprised of varying combinations of the mechanisms discussed above. Numerous interesting variations in the combinations of drought-avoiding mechanisms could be cited, but that would be beyond the scope of this brief introduction to the topic.

TOLERANCE

Drought tolerance refers to the ability of plant tissues to withstand water stress. The degree of tolerance varies widely among plant species and among growth stages within species. Tolerance ranges from the ability to maintain photosynthesis and plant growth at low cell-water potential to the ability to survive periods when tissues are air-dry (Table 11). It is convenient to divide drought tolerance into that found in lower plants (such as algae) and that of higher plant forms.

TABLE 11: Drought Tolerance Levels in Various Plant Species

Plant Group	Genus	Tolerance Level for Survival	Source
Cyanophyta	*Nostoc commune*	air-drying	Parker 1968
Lichens		air dry for:	Larcher 1975
	Lobaria pulmonaria	8 to 38 weeks	
	Peltigera canina	38 to 54 weeks	
	Umbilicina pustulata	62 to 94 weeks	
Bryophyta	*Herberta juniperina*	39% RH[1]	Parker 1968
	Bazzonia stolonifera	75% RH	
	Neesioscyphus spp.	96% RH	
Pteridophyta	*Ceterach officinarum*	0.02% RH, 5 days	Parker 1968
	Raniondia spp.	air dry	
Higher plants	*Triticum* spp.		
	seedling	98 to 99% WSD[2]	Oppenheimer 1960
	2 to 3 leaf stage	30% WSD	
	Oropetium capense	0% RH	Gaff 1971
	Helianthus annuus	52% WSD	Larcher 1975

[1] RH = relative-humidity level of plant equilibration.
[2] WSD = water saturation deficit:

$$\frac{fully\ turgid\ weight - fresh\ weight}{fully\ turgid\ weight - dry\ weight} \times 100$$

There are many examples of very drought-tolerant species among the lower plants. These species are referred to as being poikilohydrous, and they are capable of maintaining the integrity of all plant structures under extreme changes in hydration. A few examples of tolerant species within the lower plant groupings, along with indications of the mechanism of this tolerance, where known, will be given.

The blue-green algae (Cyanophyta) were probably the earliest form of plant life to invade land, and they are common soil microflora today. Many species of Cyanophyta can be air-dried and then will begin growth upon rewetting. Several species of Cyanophyta in India, for example, form on wet soil surfaces during the rainy season, become powder-dry in the dry season, and are capable of resuming growth in as little as 5 minutes when wetted (Parker 1968). The length of time that blue-green algae can remain desiccated and later resume growth can also be quite remarkable—for example, a specimen from a herbarium sheet regrew after 87 years (Lipman 1941). Studies of the protoplasm of resistant species of Cyanophyta have failed to reveal the physiological cause of such great desiccation resistance. Resistant cells have been reported to contain numerous small vacuoles, and the cytoplasm contains granular bodies of polysaccharides and DNA.

In the green algae, drought resistance has been reported in many species. Much of this resistance is a result of avoidance—such as surviving drought as resting spores—rather than tolerance. Some species may be able to survive in the vegetative form as well. In the marine algae, varying degrees of drought tolerance are exhibited, varying with the habitat of the species. Species of the intertidal zone are more tolerant than those exposed only at the ebb-tide line (Larcher 1975).

Lichens, a symbiotic combination of an alga and a fungus, also exhibit a wide range of drought tolerance. Lichens from dry sites can normally survive severe droughts, whereas those from wet sites cannot. The length of time that air-dry lichens can survive without damage was found to vary from 8 weeks in a forest species to 62 weeks in a species found on exposed rocks (Larcher 1975). Lichens in some dry habitats rely on mist or fog for moisture. Studies of the mechanisms of drought tolerance in lichens have not been extensive enough to permit general conclusions to be drawn.

A range of drought tolerance has been demonstrated in mosses (Bryophytes). Desiccation of a terrestrial moss, *Tortula ruralis*, over silica gel removed almost all water, yet upon rehydration it rapidly resumed the capacity for protein synthesis (Bewley 1973). Prolonged desiccation for up to 10 months had no adverse effects on the plant. The polyribosomes of the cells, which function in protein synthesis, were maintained during this period of desiccation. An aquatic moss, *Hygrohypnum luridum*, on the other hand, suffered irreversible loss of polyribosomes upon desiccation (Bewley 1974).

Several species of poikilohydrous ferns (Pteridophyta) have been studied in some detail. Fronds of the fern *Mohria caffrorum* can be equilibrated at 0% relative humidity without apparent injury (Gaff 1971). The tolerance of such water loss is attributed to solidification of vacuoles due to gelation of catechin-like compounds as drying progresses, which presumably helps to prevent the rupture of protoplasmic membranes (Parker 1968). Other ferns have been shown to tolerate powder dryness and still recover upon rewatering.

Extreme drought-tolerant species are not as common in higher plants as they are in lower plants. Few species can withstand water loss below about 50% of saturation water content without injury or death (Parker 1968). However, varying degrees of drought tolerance do occur. Drought tolerance is also very often associated with heat tolerance, since restricted transpirational water loss from plant surfaces can increase tissue temperatures.

Relatively few of the higher flowering plants (angiosperms) have extremely drought-tolerant mature foliage. Four poikilohydrous grasses have been reported in South Africa (Gaff 1971). Mature foliage of these "resurrection" plants can be equilibrated at 0% relative humidity and still recover upon rewatering. Detailed physiological and biochemical studies of the mechanism of this extreme tolerance have not been made. The brigalow (an Acacia tree) is capable of CO_2 assimilation at tissue water potentials as low as −50 bar (Hsiao and Acevedo 1974). Tolerance of low tissue-water potential enables this tree to keep its stomata open and thus maintain photosynthesis. Species that cannot tolerate low internal water potential must close their stomata, resulting in cessation of photosynthesis and growth. Prolonged droughts may cause such plants to lose dry weight through respiration.

The creosote bush, which is common throughout the North American deserts, is also very drought tolerant. When exposed to extreme drought, the mature leaves die (Levitt 1972). Immature leaves and buds dry out and turn brown, but they are capable of resuming growth when moisture conditions improve. Again, the mechanism of this extreme tolerance is not fully understood. The immature leaves appear to have lower resin contents and higher protein contents than the nonhardy mature leaves.

Extensive studies have been made of the drought tolerance of many crop species. Particular attention has been paid to differences among genotypes within species as part of the effort to increase yields under dry conditions. More studies of the mechanisms of drought tolerance have been carried out on crop plants than on noncrop plants.

Maize is less drought tolerant than sorghum (Beadle et al. 1973). Sorghum was able to photosynthesize at 25% of its maximum rate at low leaf-water potential (−11.5 bar), while maize had ceased photosynthesis by this point. Differences in the abilities of sorghum genotypes to survive low leaf-water potentials have also been demonstrated (Sullivan and Eastin 1974). The minimum water potential to which leaves could be stressed and still recover varied from −31 to −48 bar.

The lowering of cell osmotic potential, which is termed osmoregulation or osmotic adjustment, is a physiological mechanism that enables plants to tolerate water stress (Fig. 46). The total water potential of plants can be approximated as the sum of osmotic

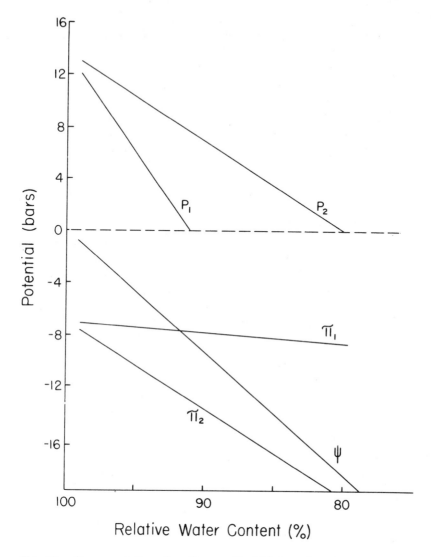

Fig. 46. Osmoregulation. Consider a cell which is undergoing water stress such that water potential (ψ) is falling. Where there is little or no osmoregulation (π_1), osmotic potential drops due to increased solute concentrations as water leaves the cell. Turger potential (P_1) becomes zero at about 88% relative water content; where osmoregulation occurs (π_2), osmotic potential drop is accelerated by net solute gain in the cell, which serves to lessen the drop in turgor potential (P_2) so that zero turgor is not reached until about 81% relative water content. Thus osmoregulation serves to counteract the effects of declining ψ on P, thereby maintaining cellular functions which depend on a positive value of P.

potential, pressure (or turgor) potential, and matric potential. Osmotic potential is a function of solute concentration in the cell, so that loss of water from the cell increases solute concentration, which in turn leads to a reduction in osmotic potential to more negative values. There is evidence that part of the reduction in osmotic·potential may be due to a net increase in cell solute concentration under stress. Numerous studies have shown that leaf osmotic potential of plants under stress falls at a faster rate than warranted by the reduction in soil water potential alone (Begg and Turner 1976). The adjustment of osmotic potential helps to maintain turgor, which is necessary for continued cell function and growth. The solutes that accumulate during osmoregulation seem to vary with plant species but may include organic acids and sugars.

As plant cells are dehydrated, the vacuole shrinks and may almost disappear but, upon rehydration, swells to its original size and shape. However, if the protoplasmic membrane is ruptured, the cell is unable to recover. For this reason species that are able to maintain cell turgor pressure and cell configuration by osmoregulation have increased tolerance of water stress.

Apart from purely mechanical injury, various cellular processes are affected by low water potential (Table 12). Cellular activities such as protein synthesis and photosynthesis exhibit varying degrees

TABLE 12: Effects of Water Stress on Cellular Processes

Stress Effect	Result
Reduced water potential, ψ	reduction in tissue growth
	limitation of hydrophytic processes
Reduced turgor pressure, π	modified growth and enzyme systems associated with the plasmalemma and cell wall
Concentration of molecules	increased activities of reactants or products alters reaction rates, particularly enzymes
Volume changes	cell volume reduced by shrinkage of plasmalemma; volume of vacuoles and cell organelles reduced
Alteration of macromolecular structures	proteins lose water, which maintains bonding within the molecule

TABLE 13: Effect of Osmotic Potential in Leaves on Photosynthetic Rate of Sorghum Varieties

| Type | μ moles O_2 evolved mg^{-1} chlorophyll h^{-1} | | |
	-5.4 bar	-11.4 bar	-31.1 bar
Milo	139.2	71.3	11.7
Shallu	131.1	86.9	50.4
Hegari	77.6	50.6	32.9
Kafir	74.1	48.6	28.6
Durra	65.5	55.7	27.2
Feterita	59.2	46.8	20.6

Source: From Sullivan and Eastin (1974), by permission of the authors.

of tolerance. Other cellular features such as protoplasm texture and amount of "bound" water have been cited as contributing to drought tolerance. However, definitive proof is lacking.

Interruption of photosynthesis cuts off the energy supply to the plant. Tolerance of the photosynthetic apparatus to water stress varies widely within a species (Table 13). Some of the lower plants can maintain photosynthesis at extremely low water potentials. A green alga (*Apatoccus labutus*) is capable of CO_2 uptake and photosynthesis at water potentials of -370 bar and possibly lower (Levitt 1972). Effects of drought stress on photosynthesis generally commence at leaf water potentials of -1 to -3 bar, and in some mesophytic plants photosynthesis ceases by -5 to -8 bar. Species that can maintain photosynthesis at low water potentials are able to continue growth during the development of drought stress (see Chap. 4).

Increased hydrolysis and reduced rate of synthesis of protein are components of drought tolerance. Extreme protein loss can cause death of the plant. The superior drought tolerance of young leaves, as opposed to older leaves, has been attributed to the higher protein content found in younger leaves, which is usually well above the minimum requirement for life (Levitt 1972).

Several theories of mechanisms of drought tolerance have been advanced. Drought injury has been attributed to mechanical damage to the cell caused by the drying and rewetting process. Early observations indicated that the vacuole shrank as the cell lost water, and that it could resume its former size and shape upon rehydration (Parker 1968). However, mechanical injury to the cell membrane, leading to rupture, prevented the vacuole from recovering. More recently the relationship between "bound" water and protein has been put forward as being important in drought tolerance. Water molecules play a part in the linkages within protein molecules, and water stress can affect protein and thus enzyme synthesis. Detailed discussions of these theories are beyond the scope of the present review but can be found elsewhere (Hsiao 1973; Levitt 1972).

Drought tolerance of plants often varies through the growth cycle. Tolerance tends to be greatest during early seedling development and decreases through later stages of development. Greater tolerance during the seedling stage assists in the establishment of the plant, particularly in arid areas. As the plant grows older, avoidance mechanisms such as extensive root systems help to offset the decline in tolerance.

Heat tolerance is a necessary accompaniment of drought tolerance under many conditions, since low tissue-water potentials can restrict transpiration rates and elevate tissue temperature. High temperatures can interrupt metabolic processes by inhibiting enzyme function or synthesis. Photosynthesis can take place over a broad range of temperatures, although the optimum rate of photosynthesis varies with species (Fig. 47). A desert herbaceous perennial, *Tidestromia oblongifolia*, reaches its maximum photosynthetic rate at a leaf temperature of $47°C$ (Bjorkman et al. 1972), while most mesophytic crop plants reach their maximum rate in the 25 to $30°C$ range.

The amount of heat buildup in plant leaves is dependent upon the amount of incident radiation and upon the ability of the leaf to reradiate heat. Transpiration is an important means of cooling leaves. Leaf color and surface texture affect reflectivity, while leaf size and shape influence heat exchange and reradiation. Plant temperature varies considerably among species. For example, wheat leaf temperatures are generally slightly below ambient air temperatures, maize leaf temperatures are slightly above ambient, while the temperature of cactus stems can be as much as $10°C$ above ambient.

The relative importance of drought avoidance and drought tolerance varies with plant species. In the lower plants, drought tolerance is by far the most important mechanism. The lichens and mosses, for example, do not possess the means to avoid drought stress since they cannot control water loss from their tissues or provide water to the stressed tissues. In higher plants, drought avoidance is generally of more importance than tolerance. Drought avoidance is of superior adaptive significance in that it allows the plant to continue growth in all but very severe droughts. Poikilohydrous drought tolerators, on the other hand, do not grow during stress periods but merely survive until moisture conditions become

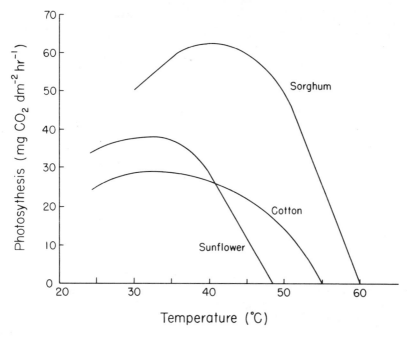

Fig. 47. The effect of temperature on photosynthetic rate of sorghum, cotton, and sunflower (after El-Sharkawy and Hesketh 1964).

more favorable. Varying combinations of drought avoidance and tolerance occur in higher plants. Plants that rely primarily on avoidance mechanisms can be found growing next to those that depend on tolerance mechanisms. The former will have closed stomata and high osmotic potential during drought stress, while the latter will have open stomata and low osmotic potential (Levitt 1972). Some degree of drought tolerance is necessary in drought avoiders, since some reduction in plant water potential is unavoidable during severe stress.

GROWTH

A great deal of the research on plant water stress has been aimed at documenting the effects of water stress on the growth of plants. Information is available on the effects of drought on processes taking place at the cellular level, on cell division and growth, on the whole plant and its components, and on plant communities or crops. In the study of the effects of drought on plant growth, it is logical to begin with the effects at the cellular level, and then progress to the plant as a whole, since the effects seen at the latter levels often have their origins in the cellular effects.

✗ As mentioned in the previous section, the maintenance of cell turgor plays an important role in cell growth. In many species, cell expansion is one of the plant processes most sensitive to drought (Hsiao 1973). Reduction in cell size is correlated with reductions in water potential of the growth medium. The elongation of *Nitella* cells is very sensitive to cell water potential changes and accelerates and decelerates rapidly as water potential is increased or decreased (Green 1968). Other evidence for the sensitivity of cell growth to drought stress can be inferred from the responses of tissues, such as leaves, to drought. Information on the role of cell turgor in growth has been provided by circumstantial evidence. There is commonly a rapid growth phase in cells or tissues following relief from short periods of stress, with the end result that there is no net reduction in total elongation. This suggests that cell metabolic events continue during stress, but cell expansion is inhibited by low turgor (Hsiao 1973). Similarly, turgor is necessary for the expansion of root cells, which must counter the opposing force of the soil. Osmoregulation can help to maintain turgor and permit cell expansion to continue, at least temporarily, under mild stress levels.

Synthesis of the cell wall, a necessary part of cell expansion, is also sensitive to water stress. Cell wall synthesis in oat (*Avena sativa*) coleoptiles is reduced by water stress and is to some extent dependent on turgor pressure (Ordin 1960). Part of the sensitivity of cell wall synthesis to water stress may be the result of reduced growth.

Cell division, a process necessary for tissue and plant growth, is considered to be less sensitive to water stress than cell enlargement (Vaadia et al. 1961). This conclusion was based on cell counts of stressed and nonstressed leaves, which showed an equal number of cells but those in the stressed leaves were smaller. The culture of radish cotyledons on media with osmotic potentials of -1 to -2 bar produced a 60% reduction in DNA content, and hence cell division, as compared to controls (Gardner and Nieman 1964). There was little further reduction if the osmotic potential of the medium was reduced to -16 bar. Other investigations have shown a similar response, while yet others have shown cell elongation to be less sensitive than cell division. The effect of water stress on cell division could indeed be partially indirect if cell division is dependent upon attainment of a threshold cell size which is sensitive to water stress (Hsiao 1973).

Following consideration of cell growth and replication, a discussion of the effects of drought stress on cellular metabolic processes is in order. It is probably logical to start with enzymes, since they are central to all metabolisms. The levels of some enzymes are reduced by water stress, while the levels of others, such as those involved in hydrolysis and degradation reactions may be increased by stress (Hsiao 1973). Water stress caused a reduction in the levels of nitrate reductase in barley, but rewatering restored the original levels (Huffaker et al. 1970). In the same study, reduction in activities of enzymes involved in photosynthesis was not as great as that of nitrate reductase. The level of α-amylase increased in vegetative tissues exposed to drought stress; on the other hand, the level decreased when germinating seeds were stressed (Hsiao 1973).

✳ Reductions in the water potential of leaves or other photosynthetic tissues reduces photosynthesis as a result of interruption in CO_2 supply by stomatal closure, or by direct effects of the water stress on the photosynthetic apparatus. Reduction in photosynthesis due to lack of CO_2 may occur on both a diurnal basis, as midday stress levels force stomatal closure, or on a more prolonged basis during severe droughts. It has been suggested that C_4 plants are better adapted to drought conditions than C_3 plants, since photosynthesis can continue at lower CO_2 concentrations. However, this is not universally true. A comparison of soybean, a C_3 plant, with maize, a C_4 plant, showed that maize photosynthesis was inhibited at higher leaf-water potentials than soybean (Boyer 1970b) (Fig. 48).

Under severe drought-stress conditions, the cellular apparatus involved in photosynthesis is affected. Chloroplast activity in sunflower is inhibited by leaf water potentials similar to those which cause stomatal closure (Keck and Boyer 1974). The inhibition of activity consists of a reduction in electron transport, followed by a reduction in photophosphorylation at lower water potentials. Water stress inhibits chlorophyll formation in etiolated leaves (Bourque and Naylor 1971), but it does not appear to be particularly significant in green leaves (Huffaker et al. 1970).

✳ The effects of water stress on photosynthesis also vary with factors such as leaf age and prior growth conditions. Mature leaves do not recover full photosynthetic capacity upon rewatering, although in some species the photosynthetic sensitivity to drought of older leaves may be less than that of younger leaves. Productivity of

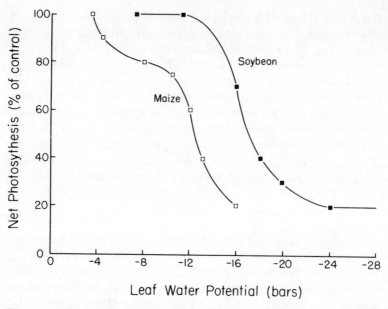

Fig. 48. Effect of leaf water potential on photosynthesis of maize and soybean (after Boyer 1970b).

stressed maize plants was found to be greater if they had been previously exposed to stress than if they had no previous exposure (Boyer and McPherson 1975).

Maintenance of photosynthetic capacity during water stress is essential to maintain growth and prevent net weight loss or death of the plant. The rate of photosynthesis must exceed the rate of respiratory loss for growth to proceed. Dark respiration is known to be affected by water stress. Rate of respiration in soybean, maize, and sunflower declines with falling leaf-water potential to about −16 bar, below which level the rate of respiration remains constant (Boyer 1970a). Other studies have shown an increase in respiration rate during the early phases of stress, followed by a reduction in rate as stress becomes more severe. Severe dehydration of tissues causes a burst of respiration in some species, possibly as a result of hydrolysis of starch to sugar. Photorespiration, which increases CO_2 production in the light, may rise briefly as stress develops and stomata close. This rise is probably due to increased leaf temperature and is only transitory in nature, since the substrates for photorespiration arise from photosynthetic products (Begg and Turner 1976).

Respirational loss of plant dry weight during severe droughts can be quite significant under certain circumstances. Some xerophytes, for example, lose dry weight during prolonged droughts due to a low photosynthetic rate. A cactus was shown to lose both water and dry weight during a 6-month drought (Walter and Stadlemann 1974).

The synthesis of protein can be affected by water stress in several ways. Protein synthesis is dependent upon a number of reactions: for example, absorption and reduction of soil nitrogen, photosynthetic production of amino acid precursors, synthesis of nucleic acids, and assembly of amino acids into proteins. Under moderate stress conditions, absorption of nitrogen by the plant is usually not a problem (Fig. 49). Indeed, the reduction in nitrate reductase activity which is observed under drought stress is not due to lack of nitrogen (Huffaker et al. 1970). Photosynthesis is essential to provide the carbon compounds which are transaminated and converted to amino acids. In citrus trees, lack of available carbo-hydrate, because of cessation of photosynthesis at relatively low stress levels, reduces amino acid levels despite adequate availability of reduced nitrogen in the plant (Naylor 1972).

The synthesis of ribonucleic acid (RNA) is relatively unaffected by moderate stress levels in older tissue, but RNA degradation by the action of the enzyme ribonuclease may increase. The synthesis of RNA in young tissue has been reported to be sensitive to moderate water stress (Bourque and Naylor 1971).

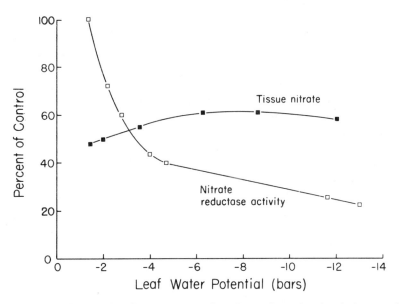

Fig. 49. Effect of leaf water potential on tissue nitrate level and nitrate reductase activity in *Zea mays* seedlings (after Morilla et al. 1973).

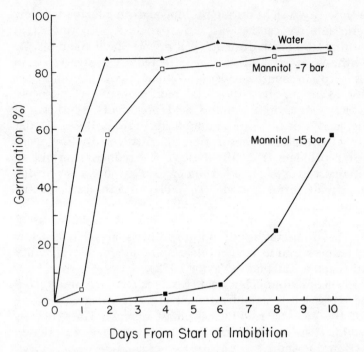

Fig. 50. Effect of osmotic potential of the germination medium on germination of alfalfa seeds (after Uhvits 1946).

The total of free amino acids in leaves increases under prolonged water stress. The most pronounced rise is that of proline. In Bermuda grass, for example, free proline in stressed plants was 10 to 125 times the value in control plants (Barnett and Naylor 1966). The accumulated proline may serve as a storage compound for reduced carbon and nitrogen during stress. The accumulation of proline has been linked with drought resistance (Chap. 4).

With this brief introduction to some of the effects of drought stress at the cellular and subcellular level, the effects of stress on the growth of entire plants and crops will be discussed. The subject will be approached on a plant phenological basis, starting with seed germination and finishing with fruit or seed maturity.

In most higher plants, the seed is the starting point for the growth cycle. Seeds begin developmental changes shortly after the start of water imbibition, provided that no other factors impose dormancy. Water uptake by seeds is dependent upon soil water content and varies with plant species. The rate of germination is slowed at low osmotic potential (Fig. 50). Various adaptations to protect germinating seedlings from drought stress have arisen. Mesophytic plant seeds, if deprived of water after seedling development has started, will die if severe drought stress is imposed. Certain desert annuals, however, have the capacity to interrupt growth after germination has started if there is insufficient moisture for continued growth (Gates 1968). *Salsola volkensii*, a summer annual of the Middle East, can start germination after a rain, but, if moisture is insufficient, the developing rootlet will not penetrate the perianth that encloses the seed. The germinating seeds can be dried to their preimbibitional water content up to 26 hours from the start of imbibition without injury. Development continues normally upon rewetting. Wheat seeds imbibe water quickly and germinate at low soil-water potentials. There is some evidence to suggest that the very early stages of growth of the seedlings are drought resistant, but this resistance is largely lost by the time of first-leaf emergence (Oppenheimer 1960).

Germination and establishment of a seedling is followed by a period of vegetative growth, which in most species overlaps the period of development and growth of reproductive structures. This vegetative growth phase is important to the plant because it supplies the assimilates for early development of reproductive structures. The sensitivity of plants to drought stress during this period varies considerably, as do the aftereffects seen in later phenological stages.

As discussed earlier, cell expansion and division can be affected by relatively small water stresses, usually before photosynthesis is affected. This has a profound effect on the expansion of leaves and stems of growing plants (Fig. 51). In maize, for example, rates of leaf enlargement are greatest when leaf water potential is in the -1.5 to -2.5 bar range (Boyer 1970a). As leaf water potential falls to -9 to -10 bar, leaf elongation virtually ceases, while photosynthesis is impaired only slightly. Sunflower leaves are even more sensitive to water deficits—growth virtually ceases at -3.5 bar (Boyer 1968). During the diurnal cycle, leaf water potential can be expected to fall below this point during the day but recover again during the night. As a result, much of the sunflower leaf growth occurs at night. When drought stress becomes more severe, the rate of elongation decreases. If the stress becomes so great that leaf water potential does not recover above -3.5 bar at night, leaf growth ceases entirely.

Water stress also affects leaf area through its effect on the initiation of new leaves. For example, in barley plants subjected to stress, initiation of new leaves ceased, while existing leaves continued to expand (Husain and Aspinall 1970). The cessation of leaf initiation

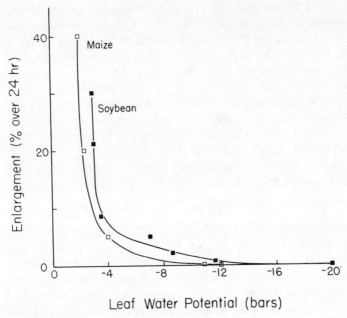

Fig. 51. Effect of leaf water potential on leaf enlargement rates in maize and soybean (after Boyer 1970a).

is probably caused by the effect of water stress on cell division and enlargement in meristematic tissue.

As might be expected, there is a considerable range among species in sensitivity of leaves to water stress. However, the general effects of stress are the same—reduction in number, rate of expansion, and final size of leaves. Xerophytes and halophytes (adapted to saline sites) can continue leaf development under somewhat lower leaf-water potentials than mesophytes.

Continued drought stress has a further effect on plant leaf area. As stress becomes more severe, leaf drop is accelerated, particularly in older leaves. This serves as a drought-avoidance mechanism in many plants, as discussed earlier. Although leaf-area reduction can help to limit plant water loss, it may also reduce photosynthetic capacity to the extent that subsequent plant development is impaired. Some of the carbohydrate and nitrogen compounds may be remobilized from the senescing leaves and transported to the seed.

Stem growth, both length and diameter, can also be inhibited by drought stress. One of the classic manifestations of drought in many plant species is a marked reduction in total plant height. In crops this may or may not be significant, depending upon whether the economic part of the crop is the vegetative or the reproductive portion. The stem has an important role in translocation of mineral nutrients, water, and assimilates within the plant. Reduction in stem height reduces the distance water moves from soil to leaf to the atmosphere without reducing the resistance to water flow.

There is little information available on the direct effects of water stress on nutrient transfer. In the tomato plant, there is some indication that transport of ions from root to shoot is inhibited by water stress (Hsiao 1973). At the same time, water stress reduces the growth of the plant as a whole, thereby reducing the demand for mineral nutrients.

The transport of photosynthetic assimilates from the source in leaves to sinks in other plant parts is reduced by water stress. The reduction is produced either by a direct effect of stress on the transport system or from an indirect effect due to stress-induced reductions in photosynthesis, and hence assimilate availability. A study of the effect of water stress on sugarcane showed that stress reduces assimilate translocation more than it reduces photosynthesis (Hartt 1967). This led to the conclusion that translocation is more sensitive to water stress than photosynthesis. In maize, water stress was also found to inhibit translocation more than photosynthesis (Brevedon and Hodges 1973). An investigation of the effect of water stress on translocation in ryegrass (*Lolium temulentum*), however, showed that translocation is less sensitive to water stress than photosynthesis (Wardlaw 1969).

The contrasting results obtained in these studies may be explained by differences in stage of plant development and effects of water stress on growth. If the drought stress occurs at a particularly sensitive developmental stage, growth will be inhibited, thus reducing assimilate demand. Consequently, inhibition of translocation would occur. On the other hand, if the stress occurred at a stage that did not affect the growth of the assimilate sink but reduced photosynthesis, no diminution of translocation would be seen until assimilate supply became limiting. Indeed, removal of leaves in stressed ryegrass, which increases the sink:source ratio and stimulates assimilate demand, caused very little reduction in translocation rate (Wardlaw 1969). The present concensus is that the effect of drought stress on translocation of assimilates is caused by reduced growth and/or photosynthesis.

Severe water stress may have an effect on water transport in the plant. As stress increases, water tension in the xylem increases, which can cause the water column in the xylem to break or cavitate

(Hsiao 1973). This increases the resistance to water flow through the xylem.

After discussing some of the effects of water stress on vegetative plant parts, the next logical step could be to look at the integrated effects of these factors on overall vegetative growth of plants. Most comparisons of the growth of stressed and nonstressed plants have been made on crops.

In addition to the factors mentioned above such as reduced leaf growth rate and general stunting of the plant, drought stress inhibits the initiation and growth of secondary tillers in plants of the grass family. This leads to reduced herbage production in the case of forage crops, or to the potential loss of seed yield in cereals. A typical response to drought stress in crop plants is a general lowering of measured growth rates in the plant as a whole or in its component parts. For example, crop growth rate (CGR) of a rape (*Brassica napus*) crop, measured as dry-matter production/unit land area, was reduced in stressed plants compared to nonstressed (irrigated) plants (Clarke and Simpson 1978) (Fig. 52a). Growth rates in the component plant parts, leaves and stems, were also reduced. The rate of leaf-area increase was slowed, and leaf-area duration was also reduced due to premature leaf-drop in stressed plants (Fig. 52b).

Periods of water stress at certain stages of crop development can be of benefit to the economic quality of the crop. In sugarcane, for example, allowing the crop to dry out near maturity produces a higher stem-sugar content at harvest (Hartt 1967).

One remaining aspect that merits discussion with regard to water stress during vegetative growth is the repair of damage accompanying the relief from stress. Alleviation of stress is followed by a rapid rise in leaf water potential and recovery of turgor. In some species there may be a lag in the full reopening of stomata, varying with leaf age and species (Glover 1959). When plants are recovering from stress there may be a brief period during which growth exceeds that of nonstressed plants. The extent of recovery depends upon the duration of stress; full recovery is generally not observed after severe stress (Boyer 1970b). The significance of stress during vegetative growth on later growth stages varies with such factors as plant species, length and severity of the stress, and subsequent environment. The sensitivity of the major growth processes to drought are summarized in Table 14.

The development of roots under drought stress is of particular significance to plants, since an extensive and efficient root system is an important drought-avoidance mechanism. In addition to the absorption of water, roots absorb mineral nutrients from the soil.

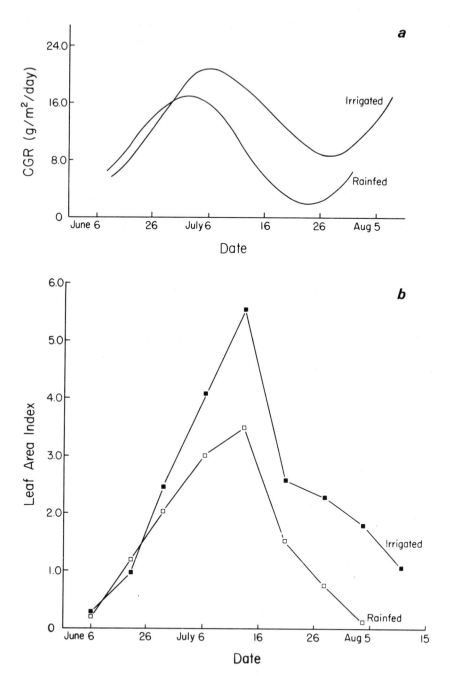

Fig. 52.   Leaf-area index and crop growth rate (CGR) of rape (*Brassica napus* L.) under rainfed and irrigated conditions.

TABLE 14: Relative Sensitivities of Plant Growth Processes to Water-Deficit Stress

Sensitivity	Process	Consequence
Most sensitive	Cell enlargement	reduction of leaf and stem growth and reproductive structures
	Cell-wall synthesis	
	Protein synthesis	reduces enzyme activity
	Stomatal opening	reduces photosynthesis
	Photosynthesis	reduces growth
Least sensitive	Translocation	limited by availability of photosynthate

Hence, water stress and plant nutrition interact. The growth of roots is also dependent upon the above-ground portion of the plant for assimilates. Thus, any conditions that impede photosynthesis in the tops will ultimately affect the roots.

Understanding of the growth of roots in relation to soil moisture has been the subject of controversy for some time. Under extreme drought conditions, root growth is inhibited if cell turgor is reduced sufficiently to stop cell elongation. In a study of wheat, oat, and barley roots, there was a progressive inhibition of root growth as water stress built up (Salim et al. 1965). There was little penetration by roots into soil below the permanent wilting point, and leaf growth tended to continue after the cessation of root growth. Two native grasses, on the other hand, appeared to be more capable of penetrating dry soil than the cereals. The roots of flax continued elongation in areas of relatively moist soil even though the plant as a whole was showing signs of severe stress (Newman 1966). The suggestion has often been made that root development is enhanced relative to shoot development during periods of water stress. There is some evidence to suggest that shoot:root ratios are lower in dry than in wet habitats, but the ratio is also affected by other factors such as soil structure and the stage of plant development. Total root-system size, however, is greater in nonstressed than in stressed plants (Fig. 53).

Rooting density of cotton decreased at soil water potentials below -1 bar (Taylor and Klepper 1974). During drying of the soil, there was also a shift in the rooting density and depth in cotton (Klepper et al. 1973). At the start of soil drying, root density was greatest near the soil surface, but after drying had progressed, root density was greater in the deeper regions of the root profile. This arose from the death of older roots near the soil surface and the growth of new roots at greater depths.

The growth of deep root systems to exploit deeply stored soil moisture when surface moisture is depleted helps plants to avoid drought. The efficiency of extraction from deep roots can be quite high, particularly when surface roots are in dry soil. This has been demonstrated in soybean (Stone et al. 1976) and maize (Taylor and Klepper 1973), where it was observed that roots deep in the profile extracted water more effectively per unit root length than roots nearer the surface. Deeper roots tended to be younger—and thus more active in absorption—and less densely packed, reducing root competition. It is evident that extensive root systems per se are of less importance than the high activity of roots in moist soil zones. Older roots tend to become suberized and, therefore, ineffective in water absorption.

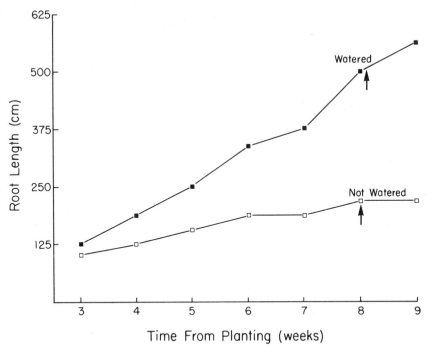

Fig. 53. Root lengths of wheat (*Triticum aestivum* cv. Thatcher) showing on glass-faced boxes for watered and unwatered soils. Arrows indicate time of heading (after Hurd 1974).

The interaction between the root and shoot is if considerable importance: roots require assimilates from the shoot, while the shoot requires water and nutrients from the root. Supply of assimilate to the root system is dependent upon the stage of plant development as well as upon environmental factors that affect photosynthesis. In annual plants, root growth is greatest during the vegetative growth phases due to the distribution pattern of assimilates. When reproductive structures begin rapid development, most assimilates are routed in their direction, which causes cessation of root growth. The shoot:root ratio tends to become higher in annuals during the vegetative growth stage (Aung 1974).

Environmental conditions, such as water stress, that restrict photosynthesis should in turn inhibit root growth. In cotton, for example, root length stopped increasing when plant top growth slowed or ceased as water stress developed (Taylor and Klepper 1974). There was an indication that shoot growth was not inhibited as early as root growth in stressed cereal seedlings (Salim et al. 1965). The timing of root- versus shoot-growth cessation is a function of shoot and root water status, which affect growth and metabolism, as well as perhaps the sink strength of shoot components as compared to the root. For example, if developing leaves are a stronger sink for assimilates than developing roots, root growth could be expected to cease before shoot growth when water stress begins to limit assimilate supply.

Aerial plant parts are dependent upon the root system for the supply of water and nutrients required to maintain growth. High plant-transpiration rates can increase ion uptake by roots when soil water is available. Conversely, at low soil-water levels, ion uptake is reduced. Reductions in nitrogen and phosphorus uptake in response to water deficits are well documented. In tomato plants, for example, uptake of nitrogen and phosphorus was reduced during water shortage; reduction of phosphorus uptake was proportionately more than nitrogen uptake (Gates 1957). The depression of ion uptake preceded the effect of the water stress on shoot growth. Reduced nutrient availability to the plant shoot can be expected to curtail growth. The relative importance of the effects of reduced nutrient availability and reduced water depends upon root depth and upon water and nutrient dispersion in the soil profile. In the case of crop plants, fertilizer nutrients are placed near the soil surface at planting. If surface water is depleted and roots are able to exploit more deeply stored water, nutrient supply may be inhibited before water availability. The water-use efficiency of plants will be higher under conditions of adequate nutrition, since available water is used more effectively.

The next major phenological stage in the plant is that of reproduction. Development of floral structures begins during the vegetative growth period, while anthesis and fertilization generally marks
the termination of vegetative growth in many plants of economic
importance. The two periods of the initiation of floral development
and of fertilization are particularly sensitive to environmental stress,
such as water deficits.

Reproductive development begins with floral initiation, that is,
the conversion of vegetative meristems to the reproductive stage and
the formation of floral initials. These growth phases are sensitive to
drought stress (Gates 1968), and damage at these stages can limit the
subsequent yield of seeds, particularly in plants with a determinate
growth habit. Indeterminate plants undergo floral initiation over a
longer time span, and floral development in periods of lower stress can
compensate for inhibited development during periods of stress.
Determinate plants, on the other hand, undergo floral initiation
over a brief time period, during which drought can reduce floral
development.

In barley, a determinate plant, water deficits inhibited the
production of primordia in the apical meristem (Husain and Aspinall
1970). The production of new leaf, or floral, primordia was inhibited
at soil water potentials that had little or no effect on growth of the
plant as a whole. Growth and development of existing primordia was
less sensitive to drought stress. Similarly, in *Lupinus albus*, the rate
of primordial initiation was quickly reduced as water stress developed
(Gates 1968). Development of foliar and floral primordia resumed
rapidly upon rewatering.

Fertilization is sensitive to drought stress. When maize plants
were subjected to water stress during the pollination period, yield
was reduced by 22% (Robins and Domingo 1953). The yield reduction
was not due to lack of pollen, since an ample supply was available
from adjacent well-watered plants. Growth of the pollen tubes was
perhaps inhibited by water stress in the styles. Peas are more sensitive
to water stress during flowering than during either the vegetative or
seed-filling phases of growth (Hiler et al. 1972). Similar responses
have been shown in rape (*Brassica napus*) (Mingeau 1974) (Fig. 54).

After fertilization, at which time the maximum number of
seeds is fixed, water stress affects the final seed number and size,
which in turn determines seed yield. Most of the drought effects
during this period arise from reductions in assimilate supply due to
reduced photosynthetic rate and shortened duration of seed growth.
Although it is often stated that the seed-filling stage seems to be less
sensitive to drought stress than the floral initiation to flowering

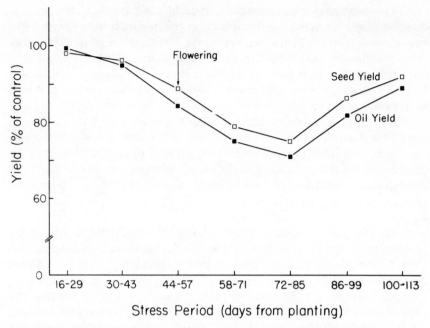

Fig. 54. Effect of time of water stress on seed and oil yield of *Brassica napus* (after Mingeau 1974).

period, there is ample evidence of the effects of prolonged drought on the reduction of yield of crops by reduced seed-filling.

In the cereals and other members of the grass family, seed number is fixed at fertilization. There is evidence that in certain other species seed number can still be reduced for some time after fertilization by the abortion of developing ovules. One plant that shows this response is rape (*Brassica napus*). The number of seeds per pod tends to be uniform throughout the plant at the end of the flowering period, but progressive seed abortion reduces the final seed number (Clarke 1979). Seed abortion appears to be caused by a lack of assimilate and is greatest in the lower regions of the plant. The reduction in seed number is most severe under drought stress, which reduces overall plant growth rate and photosynthate production.

The effect of drought stress reducing final seed size is very common. Seed size is the final component in the development of yield. Favorable moisture conditions during seed-filling can compensate to some degree for stress at earlier stages, or, conversely, drought stress can limit yield despite favorable moisture conditions earlier in development.

In cereals, most of the photosynthate used in grain-filling arises from current photosynthesis in the ear, leaves, and stem. Reductions in the assimilate supply to the grain could come from reduced photosynthetic rate and photosynthesizing area, or from reduced translocation. As noted previously, there is some controversy concerning the relative sensitivities of photosynthesis and translocation to drought stress. There is no doubt that drought stress reduces photosynthetic rate. In addition, drought can reduce assimilate availability through its effect on leaf-area production and duration. However, the question remains as to which inhibits grain-filling—lack of photosynthetic capacity, or inability to translocate assimilates to the grain. Information presently available points to reduced assimilate availability as the cause of reduced grain size, rather than interruption of the translocation process during drought stress (Wardlaw 1976).

Drought stress also acts on seed size through shortening the seed-filling period. Since seed-filling in cereals occurs during a relatively brief period, seed size can be markedly affected if the seed-filling period is reduced by only a few days. The reduction in duration of seed-filling by water stress is common in many plant species.

The overall effects of drought stress during development of reproductive structures and growth of seeds varies with species and duration of the stress period. Compensation for the reduction in one component of yield by the increase in another is a common phenomenon. If drought stress occurs early in reproductive development and causes a reduction in seed set, seeds may grow bigger and thereby moderate or eliminate the overall effect on yield, provided of course that stress during seed-filling is not too great.

Perennial plants exhibit many of the same responses to drought stress as annuals, but the perennial growth habit generates some additional differences. Herbaceous perennials that undergo a seasonal die-back of virtually all top growth behave in much the same manner as annuals, with the exception of the root systems. Nondeciduous perennials that maintain photosynthetic tissue throughout the year present a somewhat different case.

Pattern of root growth can differ considerably in perennials from that of annuals, and so it contributes to differences in response to drought stress on the part of perennials. Trees in cold climates may initiate root growth prior to onset of shoot activity, while trees in warm climates may have root growth regardless of shoot activity (Zahner 1968). Development of water deficits through the growing season in areas where seasonal growth occurs slows growth and causes premature cessation of root growth. Root growth can resume upon relief of this stress, even late in the growing season when shoot activity does not resume. The perennials, then, exhibit some buffering between shoot and root so that root activity is not tied to current photosynthetic activity in the shoot, as it is in annuals.

The seasonal pattern of growth is also somewhat different in the herbaceous perennials when compared with annuals. In the perennials, root growth tends to occur during vegetative growth. Root growth tapers off during reproductive growth and resumes after seed maturation until seasonal growth is over. The significance of this seasonality of growth has long been recognized in the management of forages, particularly legumes, in the temperate zone. In plants such as alfalfa, cutting late in the growing season is avoided, since defoliation reduces the supply of assimilates necessary for the buildup of root reserves prior to winter. Reduction in root reserves lowers the ability of the plant to survive the winter and to produce the initial foliage in spring. Severe drought stress during this critical period can produce the same effect by reducing photosynthesis.

Root systems of perennials and annuals tend to be more poorly developed in dry habitats than in habitats of more optimal moisture regime. Shoot:root ratios may be lower in dry sites if the drought stress reduces shoot growth proportionately more than root growth. Maximum depths of root penetration are greater in many perennials than in annuals by virtue of the longer growth period of the former. The depth of penetration, however, tends to depend upon the depth of moisture in the soil, as well as upon genetic factors (Mayaki et al. 1976). As mentioned previously, roots of some members of the cactus family are widely dispersed near the soil surface (Walter and Stadlemann 1974). When the soil is wet from precipitation, lateral root growth is initiated rapidly in order to absorb all available water.

Vegetative growth of perennials is subjected to the same basic constraints that affect annuals: for example, constraints imposed by low turgor and reduced photosynthetic rate. However, drought during parts of the season can greatly influence growth in the following growing season in perennials. For example, in tree species in which the entire leaf canopy arises from primordia in buds overwintered from the previous year, drought stress during bud formation can limit leaf number the following year (Zahner 1968).

Since perennials do not rely on seed production each year for survival, those that are very drought-resistant can survive several consecutive years of drought. Perennial trees and shrubs of arid areas, for example, have a low proportion of living tissue and can survive with very small annual-growth increments.

The effects of drought on reproductive activity in perennials can be analogous to those of annuals. For example, fruit yield and quality of domestic fruit trees is highly sensitive to available moisture during fruit formation and enlargement. In trees of temperate regions which form floral buds in the fall and flower the following spring, droughts of the previous year affect the current-year seed production (Zahner 1968). Similar effects occur in temperate herbaceous perennials. The bromegrass species *Bromus pumpellianus*, for example, undergoes floral initiation in the fall (Clarke and Elliott 1974), so that water stress during the fall initiation period influences seed production in the following year. In warmer climates, reproductive growth is adjusted to seasonal rainy periods. This is the case with many species of the Mediterranean region, which utilize the winter rainy season for this purpose.

The integrated effects of drought stress on the growth of plants have both economic and ecological significance. As we have seen, both timing and duration of drought stress are important in determining the effects of the stress. Droughts of brief duration may have relatively little long-term significance, except in the case of plant species with very low drought resistance.

Economic effects of drought vary with species. In cereals, where economic yield is the seed, the effects of moisture regime during floral development, flowering, and seed-filling have the most influence on yield. For plants utilized as forages, drought stress that affects vegetative growth of leaves and stems is of major importance. Lack of water is the major constraint to agricultural production in semiarid and arid regions. Where it is not possible to provide irrigation, the choice of drought-resistant species, the breeding of drought-resistant genotypes, and the use of the most efficient management practices helps to maximize the efficiency of the use of available water.

In native plant communities, drought stress imposes limits to the distribution of species. Only those species capable of germination, growth, and reproduction under drought stress or during brief rainy seasons can successfully invade arid habitats. The productivity of such sites, in terms of biomass per unit land surface, is extremely low when compared to moist temperate or tropical locations (Table 15).

HORMONAL CHANGES

Introduction

When Wright (1969) was examining the effect of photoperiod on the levels of abscisic acid (ABA) in wheat leaves, he observed that light

TABLE 15: Plant Biomass and Annual Growth Rate Ranges for World Vegetation Zones

Vegetation Zone	Mature Biomass $(kg\ m^{-2})$	Growth Rate $(g\ m^{-2}\ yr^{-1})$
Tropical rain forest	44 to 75	1,000 to 3,500
Rain green forest	42	600 to 3,500
Warm temperate mixed forest	24	600 to 2,500
Boreal forest	19 to 52	700 to 1,500
Tropical grassland	5	200 to 2,000
Temperate grassland	3	100 to 1,500
Desert scrub	0.1 to 4	10 to 250
Dry desert	0	0 to 10
Cultivated annual crops	3.5	100 to 4,000

Source: After Lieth (1975) and Walter (1973).

increased the level of ABA in excised leaves but had no effect on leaves of intact plants. From subsequent experiments he deduced that the large increase in ABA was not due to a direct effect of the light but to the wilting of the excised leaves. Thus ABA, previously associated with growth inhibition, was also found to be associated with water-deficit stress (Wright and Hiron 1969). This and an earlier result, demonstrating changes in cytokinin concentration in xylem exudate of water-stressed sunflower plants (Itai and Vaadia 1965), led hormone physiologists to examine water-deficit stress in terms of changes in endogenous hormones.

Plants have various responses to water-deficit stress, depending on the type, the timing, the degree, and the period of stress. The initial responses involve changes in internal water status. These changes include decreases in cellular water potential (ψ_w), solute potential (ψ_s), and turgor pressure (ψ_p). Subsequent responses include decreases in stomatal aperture, photosynthesis, leaf length, root growth, and seed set; increases in leaf senescence, abscission, and inhibition of flowering; and a lengthening (drought tolerance) or shortening (drought avoidance) of life cycle (Levitt 1972). Hormones have been implicated as controlling factors in many of these processes.

In the following sections, each hormone will be examined in turn, although it must be appreciated that in many cases hormone interactions occur. Two areas of multihormone involvement, water uptake and drought resistance, will be dealt with under separate headings.

Abscisic Acid

ABA is an inhibitor of plant growth and development. In particular, it is associated with bud dormancy, inhibition of flowering, inhibition of seed germination, and root geotropism (Walton 1980). On the cellular scale it is inhibiting to the synthesis of certain proteins (for example, α-amylase in barley endosperm) either directly or through its effect on nucleic acid synthesis (Ho and Varner 1974; Jacobsen 1977).

ABA is also associated with water-deficit stress. In leaves, the concentration of ABA rises as leaf ψ_w declines. There is a gradual rise in ABA levels (Wright 1977; Kannangara et al. 1981; Durley et al. 1981) until a critical value of leaf ψ_w is reached, at which point ABA levels increase rapidly (Zabadal 1974). This critical value was found to be −8 to −10 bar for excised leaves of corn and sorghum (Beardsell and Cohen 1975), −7 to −9 bar for bean leaves (Walton et al. 1977), and −10 to −12 bar for Douglas fir needles (Blake and Ferrell 1977). Pierce and Raschke (1980) proposed that the critical value of leaf ψ_w corresponds to a zero or near zero leaf ψ_p. However, the critical value can decrease if the plant is subjected to successive stresses, due to osmotic adjustment, that is, lowering of the leaf osmotic potential (ψ_s) (Pierce and Raschke 1980).

ABA is largely synthesized in the plastids (Railton et al. 1974; Loveys 1977), although other sites of synthesis cannot be ruled out (Walton 1980). During water-deficit stress the chloroplast membrane becomes more permeable to ABA (Mansfield et al. 1978; Heilmann et al. 1980) and ABA "leaks out," thus increasing its concentration in the cytoplasm. Since the release of ABA from the chloroplast in response to water deficit is rapid, it has been suggested that ABA is initially released from a conjugated form (Mansfield et al. 1978) and this is followed by rapid de novo synthesis (Milborrow and Robinson 1973).

What is the effect of increased levels of ABA? One important effect is on the stomatal complex. Stomatal aperture is regulated by turgor changes in guard cells (Raschke 1975a, 1977). High turgor is associated with open stomata. These turgor changes are regulated by movement of K^+, H^+, CL^- ions and organic anions, particularly malate. The turgor is maintained by imported K^+ ions together with imported CL^- ions and malate anions, the latter being synthesized in guard cells during stomatal opening. The association of ABA with stomatal movements was first implied by Tal and Imber (1970, 1971a), who observed that in leaves of a wilty tomato mutant, *Flacca*, which was able to keep its stomates open under water-deficit stress, there were low levels of ABA; furthermore, the tendency of the mutant to wilt was overcome by application of ABA. It is now believed that in water-stressed plants, ABA is synthesized and released from mesophyll chloroplasts and travels to the guard cells (Loveys 1977), where it (1) inhibits K^+ uptake (Mansfield and Jones 1971; Weyers and Hillman 1980), (2) inhibits H^+ release (Raschke 1977), and (3) promotes leakage of malate from the guard cells (Dittrich and Raschke 1977; Van Kirk and Raschke 1978). These effects cause guard cells to lose turgor. Lost turgor decreases stomatal aperture. Raschke (1977) has proposed that ABA inhibition of solute uptake and promotion of solute release can be explained solely by its inhibiting the H^+/K^+ exchange process across the plasmalemma membranes of guard cells. Other cells of the stomatal complex are required to act as sinks for solutes leaking from the guard cells (Itai and Meidner 1978). There are conflicting reports as to whether CO_2, which also closes stomata, and ABA interact (Raschke 1975; Dubbe et al. 1978) or act independently (Mansfield 1976a; Itai and Meidner 1978).

In some cases, total ABA levels are not correlated with stomatal conductance. This may be due to compartmentalization of ABA, one compartment being connected to the guard cells while the other is not (Raschke et al. 1976; Raschke and Zeevaart 1976). During water stress, it was observed that stomates closed before rises in ABA levels could be detected. To explain this, Beardsell and Cohen (1975) suggest that the initial response to stress was a redistribution of ABA, while Walton et al. (1977) suggest that the initial rise in ABA level was too small to detect. Furthermore, the stomatal apparatus may be regulated by the rate of synthesis of ABA which, due to degradative processes, may not be related to a rise in ABA levels (Walton et al. 1977).

When water-deficit stress is removed, leaf stomata open and ABA levels decline. Generally, an inverse relationship between ABA levels and stomatal opening is observed (Beardsell and Cohen 1975). In some plants, stomata do not reopen immediately after leaf turgor is restored (Beardsell and Cohen 1975; Dörffling et al. 1977; Weyers and Hillman 1979) and this "aftereffect" is attributed to residual ABA or its metabolite, phaseic acid (PA) (Walton 1980). In some cases stomata open before total leaf ABA has declined, again implying compartmentalization of ABA. For example, in *Hordeum vulgare* leaves, Cummins (1973) showed that although total ABA level remained high after removal of stress, stomata opened when ABA was no longer in the transpiration stream.

When plants are subjected to repeated cycles of water-stress treatments, the leaves are able to withstand lower leaf-water potentials before stomatal closure (McCree 1974; Brown et al. 1976) due to osmotic adaptation (Brown et al. 1976; Cutler and Rains 1978; Jones and Turner 1978). This is particularly important under field conditions, where plants encounter frequent stress periods (Davies 1977; Ackerson et al. 1980; Kannangara et al. 1981). Davies (1978) observed that stomatal sensitivity in *Vicia faba* leaves to exogenously applied ABA increased as a result of water-stress pretreatments. Furthermore, Ackerson (1980) found that the level of ABA increased in fully turgid cotton leaves due to water-stress pretreatments and suggested that the increased stomatal sensitivity to ABA was the result of prestressed accumulation of ABA. The high levels of ABA found in prestressed plants may be associated with the plants' adaptation to stress.

Besides the action of ABA on the stomatal complex, the increased level of ABA during water-deficit stress may inhibit some metabolic processes, which then lowers the demand for water in the leaf. In *Avena* coleoptiles both ABA and water-deficit stress inhibited the rate of protein synthesis (Dhindsa and Cleland 1975). The protein pattern was not the same in the two treatments, but it is possible that water stress influences protein synthesis by other means as well as through ABA.

A large increase in ABA was observed in the phloem of lupin plants subjected to water stress (Hoad 1978). It was suggested that ABA was actively transported out of mature leaves in the phloem and was able to accumulate in other parts of the plant such as pods and seeds. This accumulation of ABA may be responsible for stress-induced inhibition of flowering and reductions in both seed set and growth. King and Evans (1977) observed that a brief water stress during induction of flowering in *Lolium temulentum* was sufficient to reduce flowering response, and they suggested that this inhibition to flowering was a result of ABA accumulation at the shoot apex. Morgan (1980) indicated that reduced seed set in wheat during water stress was due to enhanced ABA levels in the spikelet.

Phaseic and 4′-Dihydrophaseic Acids

When ABA levels increase in leaves following water-deficit stress, so do its metabolites, PA and 4′-dihydrophaseic acid (DPA) (Harrison and Walton 1975). The effect of PA on stomatal conductance is much less than that of ABA, and DPA has no effect (Sharkey and Raschke 1980). The "aftereffect" of stress (continued stomatal closure after leaf turgor is restored) could therefore be caused by residual PA, but the theory needs to be tested. Kriedemann et al. (1975) proposed that the impairment of photosynthetic apparatus, frequently observed following water-deficit stress, could be caused by endogenous PA since plant extracts containing PA inhibited photosynthesis. Recent evidence has indicated that this inhibition was due to solvent impurities in the extracted PA (Sharkey and Raschke 1980).

Farnesol

In addition to ABA, there is another sesquiterpenoid antitranspirant. All-*trans*-farnesol is found in leaves of water-stressed sorghum plants (Wellburn et al. 1974) and when reapplied to the leaves of nonstressed sorghum plants it closes stomata in a manner comparable to that of ABA (Fenton et al. 1977). It is unlikely that it exerts its effect by conversion to ABA (Wellburn et al. 1974). However it is not known if farnesol can act directly on stomatal movements, although it has been proposed that short-term stomatal responses in sorghum could be attributed to farnesol (Fenton et al. 1977). Mansfield et al. (1978) have suggested that farnesol may alter the permeability of the chloroplasts to ABA, allowing ABA to enter the cytoplasm. This implies that farnesol may regulate the availability of ABA.

Xanthoxin

A number of naturally occurring compounds, similar in structure to ABA, are found to be antitranspirants. One of these compounds is xanthoxin, a growth inhibitor, which is about half as active as ABA in closing stomata of detached leaves (Raschke et al. 1975). However, at this time it is difficult to propose a role for xanthoxin since water-deficit stress does not increase xanthoxin levels (Zeevaart 1974).

Cytokinins

Cytokinins are purine derivatives that cause cell division and retard leaf aging. A major site of synthesis is in roots. From there, cytokinins can be transported to the shoots via the transpiration stream (Kende 1965; Short and Torrey 1972; Wareing et al. 1977). When Itai and Vaadia (1965) subjected sunflower plants to water-deficit stress, they found that cytokinin concentration in the root exudate was markedly reduced. Even if stress was applied to leaves by passing air over them, the root exudate still contained reduced cytokinin concentration (Itai and Vaadia 1971). In detached lettuce leaves, water stress had the effect of decreasing cytokinin levels, as ABA levels increased (Aharoni et al. 1977). After recovery from water stress, cytokinin levels increased and ABA levels decreased. The increase in leaf cytokinins in the absence of roots implies either synthesis in the leaves or a type of inactivation by water stress that is reversed when stress is removed (Itai and Vaadia 1971; Aharoni et al. 1977).

Applied cytokinin (kinetin) has the effect of reducing ABA levels in water-stressed leaves (Bengtson et al. 1979) and increasing leaf transpiration (Livne and Vaadia 1965; Incoll and Whitelam 1977; Meidner 1967; Luke and Freeman 1968; Mittelheuser and Van Steveninck 1969; Mizrahi et al. 1970; Pallas and Box 1970; Copper et al. 1972; Aharoni et al. 1977; Bengtson et al. 1979). Furthermore, it has been shown that kinetin increases the uptake of K^+ by guard cells (Van Steveninck 1972). The above effects of applied kinetin are somewhat variable, but this may be due to changes in endogenous levels of cytokinins and the physiological status of the tissue (Livne and Vaadia 1972). Young barley leaves, for example, which have high levels of endogenous cytokinins, had to be excised and kept for 24 hours before stimulation of transpiration by applied kinetin could be observed (Livne and Vaadia 1965).

In general, cytokinins and ABA have opposite effects on stomatal conductance (Bengtson et al. 1979) and concentrations are recipro-cally related in water-stressed leaves. Cytokinin levels may increase in some plants as they adapt to water stress. In excised tomato cotyledons stressed by mannitol solutions, cytokinin levels decreased but later recovered to levels higher than controls (Ong 1978).

Gibberellins

There is little information concerning the role of gibberellins (GAs) in the course of plant adaptation to water deficit stress. In excised stressed barley leaves, applied GA_3 increased the transpiration rate, an effect similar to that observed for kinetin (Livne and Vaadia 1965). However, in excised oat leaves, GA_3 was ineffective (Luke and

Freeman 1967). In excised lettuce leaves subjected to water stress, endogenous GA_3 levels were reduced, declining before ABA levels were observed to rise and at a rate that positively correlated with leaf-water saturation deficit (Aharoni et al. 1977). GA levels also declined in tomato roots, shoots, and bleeding sap following water stress applied by flooding roots (Reid and Crozier 1971).

Gibberellins and cytokinins apparently act in a similar way during water-deficit stress. It is possible that the reduced amount of cytokinins synthesized in the roots and transported to the shoots during water stress has the effect of reducing GA levels in roots and shoots (Railton and Reid 1973).

Applied GA_3 has been shown to promote leaf abscission (for example, see Chatterjee and Leopold 1964), especially during water stress (Perween and Ahmed 1976) or in the presence of ethylene (Morgan and Durham 1975). Whether GA plays a natural role in this process is not known.

Auxins

The natural auxin, 3-indolylacetic acid (IAA), has no effect on leaf transpiration (Johansen 1954); and although synthetic auxins such as 2-naphthoxyacetic acid, 1-naphthylacetic acid, and 2,4-dichloro-phenoxyacetic acid (2,4-D) were effective in stomatal closure, they were applied in high concentrations (1 mM), well above the normal physiological range (Mansfield 1967). IAA levels increased in both waterlogged or droughted broad bean plants, and this was related to ethylene levels (Hall et al. 1977).

The above results seem to suggest that high auxin levels in shoots are associated with stomatal closure and water stress. However, the opposite is implied by the following results. Tal and Imber (1970) compared shoots of the wilty tomato mutant *Flacca*, which as previously described keeps its stomates open under water stress, to a control variety and found high auxin levels associated with low ABA levels in the mutant. Furthermore, prolonged treatment of tomato plants with low concentrations of 2,4-D increased transpiration rate (Tal and Imber 1971b). Darbyshire (1971) found increasing IAA-oxidase activity in shoots of mannitol-stressed pea and tomato plants with increasing water deficits, implying that IAA levels were reduced by water stress. Also, pretreatment of maize and winter wheat seeds with IAA reduced the level of IAA-oxidase activity in the roots and shoots of 7-day-old plants under mannitol stress (Nowakowski 1979). Kannangara et al. (1981) found that free IAA levels in sorghum leaves decreased as a result of water-deficit stress. However, under these conditions conjugated IAA (IAA released by alkali treatment) sometimes increased, implying conversion to a storage form (Bandurski 1979) of the hormone during water stress.

There is now considerable evidence to suggest that ethylene mediates leaf abscission (see below). It is suggested that the action of ethylene is mediated by an inhibition of auxin transport in the leaf petiole (Beyer and Morgan 1971; Beyer 1973). Exogeneous IAA, applied to debladed petioles of explants (Craker et al. 1970; Louie and Addicott 1970) or seedlings (Perween and Ahmad 1976) delayed abscission, presumably by replacing the natural auxin synthesized and transported from the leaf blade (Wetmore and Jacobs 1953).

Since leaf senescence and abscission can occur as a result of water stress, ethylene levels and auxin transport capacity have been measured in petioles during water-stress treatments. In cotton plants subjected to water stress, not only did ethylene levels in petioles increase (McMichael et al. 1972) but auxin transport capacity declined (Davenport et al. 1977; Morgan et al. 1977). Furthermore, auxin transport capacity declined with leaf age (Davenport et al. 1980), which correlates with the observation that older leaves are the first to abscise during water stress (McMichael et al. 1973).

The decreased (or increased) levels of IAA in water-stressed plants may affect leaf and stem elongation. Since IAA-induced elongation of cells is dependent on cell osmotic potential (Kazama and Katsumi 1978), it would therefore be likely that water stress would directly affect the action of IAA in shoots. Auxins can affect water uptake by roots, and this will be discussed later.

Ethylene

Abscission is known to involve enzymatic dissolution of cell walls in the abscission zone, with the result that leaves, flowers, and fruits are cast off the plant. Ethylene is known to have a primary role in this process (Jackson and Osborne 1970) by promoting the synthesis (Horton and Osborne 1967) and movement (Abeles and Leather 1971) of hydrolytic enzymes such as cellulase, which acts on cell walls. In cotton plants, water-stress-induced abscission of leaves (McMichael et al. 1973) is promoted by ethylene (Jordan et al. 1972). It has also been shown that in water-stressed cotton plants (McMichael et al. 1972) and broad bean plants (El-Beltagy and Hall 1974) ethylene levels are rapidly elevated in both leaves and petioles.

Ethylene levels have been observed to increase rapidly in wilting detached orange leaves (Ben-Yehoshua and Aloni 1974) and wheat leaves (Wright 1977). In detached wheat leaves, stress-induced ethylene levels are enhanced by pretreatment with benzyladenine, IAA, or GA (presumably by retarding aging) and are inhibited by ABA (Wright 1979, 1980). Wright suggested that since ABA levels are also enhanced by water stress, ABA may regulate excessive ethylene production. Ethylene levels are also known to increase in roots following water-logging (Kawase 1972, 1974; El-Beltagy and Hall 1974; Jackson and Cambell 1976).

During water stress, ethylene plays an important role in the processes of abscission of leaves, flowers, and fruits (Hall et al. 1977). Since ethylene can inhibit cell extensibility, it may also be involved in water-stress-induced inhibition of root and leaf growth (Hall et al. 1977). Even though ethylene levels are observed to increase before ABA levels increase, ethylene is not believed to regulate ABA levels in water-stressed leaves (Wright 1977).

Effects of Water Uptake and Ion Transport in Roots

Besides controlling water loss through leaf transpiration, hormones affect water permeability and ion transport in roots. It is possible that under conditions of water stress, hormones could promote root permeability to water.

ABA levels are known to rise in roots of stressed plants (Milborrow and Robinson 1973), due possibly to transport from leaves (Hoad 1975) or alternatively to synthesis in roots (Walton et al. 1976). ABA increases water permeability into roots (Tal and Imber 1971a; Glinka 1973, 1977, 1980; Collins and Kerrigan 1974; Karmoker and Van Steveninck 1978). The increase in water uptake may be due to a direct effect of ABA on hydraulic conductivity (Glinka 1973, 1980; Collins and Kerrigan 1974), or it may be a result of ABA stimulation of ion transport into the xylem (Karmoker and Van Steveninck 1978).

During drought stress it may be advantageous for plants to increase ion uptake for osmoregulation to help maintain leaf turgor. However, the effects of ABA on ion transport into the xylem are unclear. Both promotive (Glinka 1973, 1977, 1980; Collins and Kerrigan 1974; Karmoker and Van Steveninck 1978) and inhibitory (Cram and Pitman 1972; Pitman and Wellfare 1978) effects have been reported, depending on concentration, plant species, growing conditions, and temperature (Pitman et al. 1974b). Differences in ion transport were also reported depending on whether intact or excised roots were examined (Karmoker and Van Steveninck 1979). Pitman et al. (1974a) suggested that an observed inhibition of ion export from roots to shoots as an aftereffect of water stress in barley plants may have been due to increased root levels of ABA, transported from leaves.

Cytokinin levels probably decrease in water-stressed roots, since during stress their concentration in xylem exudate decreases (Itai and Vaadia 1965, 1971). Applied kinetin has an inhibitory effect on water uptake and ion transport in roots (Tal and Imber 1971a; Collins and Kerrigan 1974; Glinka 1980).

Measurements of transpiration have been used to determine the effects of auxins, in high concentration $(1mM)$, on water uptake by barley roots (Allerup 1964). After an initial increase in transpiration, there was a gradual decline. Kozinka (1967) reported rapid inhibition of water uptake in roots after application of high concentrations of auxins. In rice roots, synthetic auxin 2,4-D initially stimulated, then diminished, Ca^{2+}-promoted K^+ uptake (Erdei et al. 1979). From the limited information available, auxin therefore appears to be inhibitory to water uptake by roots.

During water stress, the observed increase in ABA levels and the concomitant decline in cytokinin levels could promote uptake of water into roots. It is not yet clear if ion uptake is stimulated or inhibited by these hormone changes.

Growth Regulators and Breeding for Drought Resistance

There is considerable interest in breeding for drought-resistant lines of crop plants (Hurd 1974). However, it is difficult to find parameters that reliably predict drought resistance. Drought is often variable both in its severity and timing. Its effect on yield, the most commonly used selection index, is a result of a complex series of events involving interaction with various environmental stresses.

Since hormones are involved in many water-stress-sensitive processes, it is probable that they have a major role in plant drought-resistance mechanisms. Thus it may be possible to measure a spectrum of hormones as an index for selection of drought-resistant genotypes (Simpson et al. 1979).

For simplicity, breeders recognize two types of drought resistance: (1) drought avoidance, which is associated with high leaf ψ_w, increased root growth, and speeding up of the life cycle under drought conditions, and (2) drought tolerance, which is associated with plants that withstand drought by lowering leaf ψ_w and slowing down the life cycle (Simpson et al. 1979). Larque-Saavedra and Wain (1976) found that, in both maize and sorghum cultivars differing in drought-resistance qualities, ABA was accumulated in excised leaves to a greater extent in the resistant (presumably drought-tolerant) lines. Quarrie and Jones (1979) found no variation in leaves of non-stressed wheat genotypes, but during water stress they found that the ABA concentration increase was correlated with the change in leaf ψ_w for each genotype. They attributed the low ABA concentration in two of the cultivars examined to the superior drought-avoidance characteristics of those cultivars.

Stout et al. (1978) examined the behavior of two sorghum cultivars under drought-stress conditions and concluded that one, M-35, was a drought tolerator, while the other, NK 300, was a drought avoider. Kannangara et al. (1981) examined the levels of ABA, PA, and IAA in leaves throughout the life cycle of these two cultivars, which were field-grown under droughted or irrigated conditions. The different responses of the two cultivars to drought stress were reflected by changes in leaf hormone levels. Thus a high level of ABA was found in the drought-tolerant cultivar M-35, and this was associated with maintenance of low leaf ψ_w and ψ_s and with high leaf senescence. On the other hand, in the drought-tolerant cultivar NK 300, higher levels of IAA and PA were observed, and these were associated with higher leaf ψ_w and ψ_s. In another study with 10 cultivars of sorghum, free IAA was found to decrease under drought conditions, but conjugated IAA was often found to increase, implying conversion to this storage form of IAA (Kannangara et al. 1981). The two major cytokinins in sorghum leaves are zeatin and zeatin riboside (Kannangara et al. 1978), both of which decrease under drought stress (Durley et al. 1981). Comparison of the two cultivars M-35 and NK 300 indicated that whereas zeatin riboside levels were similar in both, zeatin levels were consistently higher in the drought avoider, NK 300.

Thus there are indications that ABA, PA, IAA, zeatin, and zeatin riboside could be used as indices to determine drought resistance in crop plants. There are, however, two problems to overcome. Due to low levels of hormones present in leaves, complex methods are required for their analysis (for example, see Kannangara et al. 1978; Durley et al. 1981). These methods require further refinement and automation in order to accomodate the large number of analyses required in a breeding program. Also, timing of harvests for examination of hormones must be carefully considered. There is a considerable diurnal fluctuation in the level of ABA in cotton leaves (McMichael and Hanny 1977) and the levels of ABA and PA in sorghum leaves (Durley et al. 1981), especially under drought-stress conditions, and these were negatively correlated with leaf ψ_w. Free IAA levels were positively correlated with ψ_L, but changed much less.

Summary

Many plant responses to water-deficit stress or osmotic stress are mediated through changes in the level of endogenous hormones, via changes in synthesis, release, and transport in both roots and shoots. ABA reduces stomatal aperture under water-stress conditions, thus preserving leaf turgor. This hormone may also influence other water-stress-related responses such as reduced metabolic rate, inhibition of flowering, and reduced seed set. Cytokinins and gibberellins act oppositely to ABA and have the effect of increasing stomatal aperture. Changes in the level of these hormones may affect the level of ABA, or vice versa. Free auxins decline under water-stress conditions and are associated with increased stomatal aperture, although opposite effects have been reported. A major factor in leaf abscission is ethylene, which increases in petioles during water stress. The action of ethylene may be mediated by an inhibition of auxin transport in the petiole. Ethylene may also reduce cell extensibility during water stress. Increases in ABA and decreases in cytokinin levels in roots of stressed plants may increase water permeability.

Although hormones are clearly implicated in the responses of plants to water stress, the mechanism of their action is poorly understood. Knowledge of these mechanisms is needed to understand fully the role of hormones during water stress. Furthermore, changes in interactions of hormones may be more important than the change in level or transport of a single hormone. A measurement of a spectrum of hormones under various water-stress treatments may eventually be used as an index for selection of drought-resistant genotypes.

J. M. CLARKE
A. J. KARAMANOS
G. M. SIMPSON

4

Case Examples of Research Progress in Drought-Stress Physiology

This chapter deals with three examples of globally important crop plants—wheat, sorghum, and faba beans—that have been studied in considerable depth for physiological explanations of their drought-resistance behavior. These crops illustrate the variety of approaches and different levels of understanding attained to this date. While the general response of plants to water stress may be implied from what can be seen as common responses for each crop, it can be readily seen that each crop also has responses peculiar to the species and to the environment in which it is grown.

WHEAT

Introduction

Cultivated wheat, *Triticum* spp., originated in the Middle East. Most wheats now grown are either *T. aestivum* (common wheat),

The section on Wheat was contributed by J. M. Clarke; that on Sorghum, by G. M. Simpson; and that on *Vicia Faba* by A. J. Karamanos.

T. turgidum var. durum (durum wheat), or *T. compactum* (club wheat). Wheat is grown virtually all over the world, although its relative importance in comparison to other crops varies from region to region. Currently, in terms of world acreage and production wheat ranks as the most important cereal crop: it was grown on about 228 million hectares in 1979. From its original semiarid habitat, wheat has spread to other habitats and temperature regimes ranging from temperate to tropical. Most wheat is grown in semiarid (250 to 500 mm annual precipitation) and sub-humid (500 to 760 mm) regions (Patterson 1965); there is also some production in arid regions with as little as 200 mm precipitation. Since wheat is poorly adapted to high temperatures and humidities, in the tropics and subtropics it is grown at high altitudes or during the cooler part of the year.

Yields vary widely from one producing region to another (Table 16), principally due to differences in available moisture and evapotranspiration ratios. They are highest in western Europe, and lowest in the arid and semiarid regions of production, such as parts of Australia, India, the Soviet Union, and North America. Although irrigation is an obvious means of increasing wheat yields in arid and semiarid areas, lack of water for irrigation and economic considerations generally restrict the amount of irrigated wheat. The effect of drought on wheat yields can be demonstrated by comparing yields under dryland and irrigated conditions (Table 17). In semiarid regions, yields under irrigation are generally 2 to 5 times as great as on dryland.

Research on the effects of drought stress on wheat, which has the ultimate objective of improving yields under stress conditions,

TABLE 16: Global Wheat Yields (1968–77)

Country	Yield (tonnes ha^{-1})
West Germany	4.31
France	4.02
United States	2.05
Canada	1.76
Soviet Union	1.44
Argentina	1.42
China	1.32
India	1.28
Australia	1.21
Turkey	1.20

TABLE 17: Yields of Rainfed and Irrigated Wheat

	Yield (tonnes ha^{-1})	
	Rainfed	Irrigated
Canada	1.4	3.0
Libya	0.6	2.3
Afghanistan	0.5	1.2
Israel	0.8	4.2

has taken two major approaches: optimization of the plant environment, and genetic improvement of the plant. Optimization of the environment involves such approaches as soil cultural practices, improvement of soil fertility, timely dates of planting, rates and methods of seeding, and insect and disease control. Genetic improvement of the plant through breeding is focused on improvement of yields, sometimes within relatively rigid grain-quality restrictions. Studies of plant and crop physiology play a role in management, research, and breeding. This discussion will deal firstly with physiological research on the effects of drought stress on wheat, followed by research in the management area, and, finally, a look at the achievements and future prospects of wheat breeding for drought resistance.

Physiology

Most physiological studies have had the objective of determining mechanisms of drought resistance in the wheat plant, which involve both physiological and morphological systems. Investigations have spanned a broad range of topics, from the effects of drought on specific enzyme systems to identification of physical features, such as awns, which may play a role in the reaction of the plant to drought. The field of research will be reviewed on a phenological basis, beginning with the seedling stage.

Although wheat crops are usually planted when soil moisture conditions are favorable, there are certain areas of production where soil moisture levels may be quite low at planting, or where soil salinity reduces water availability. One such area is the northwestern region of the United States, where winter wheat is frequently planted at marginal soil moisture levels. In the case of winter wheat in this area, it is imperative that the seedlings become established in the fall in order to ensure a crop in the following year. During the initial stages of water imbibition when seeds are placed in soil with a low water potential, imbibition can take place because of the extremely low osmotic potential of the dry seed. However, as imbibition proceeds and the seed osmotic potential rises, a point is reached at which no more water is absorbed. If this point is reached before the threshold moisture content for germination is attained, the seed will not germinate. The germination and emergence rate of wheat seeds is reduced as soil water potential is reduced (Helmerick and Pfeiffer 1954). Percentage germination tends to be relatively stable until the threshold water potential level for germination is approached.

Varietal differences in the rates and percentages of germination and emergence of wheat under water-stress conditions have been identified. Winter wheat breeding lines screened in the laboratory using soil at water potentials of -2.2 to -14.4 bar revealed a broad range of emergence times within each water potential level (Gul and Allan 1976a). The time required for emergence nearly doubled for each decrease in water potential of -4 bar. Under field conditions, rate of emergence and total plant stand of the same lines were positively correlated with the seedling vigor characteristics, coleoptile length, and seedling height (Gul and Allan 1976b). Another approach to screening wheat varieties or lines for their ability to germinate at low water potentials involves the use of osmotica such as salts or sugar polymers in the germination medium (Fig. 55). Studies of this nature reveal that wheat germination is severely reduced when water potential of the germination medium falls to -15 to -18 bar (Ashraf and Abu-Shakra 1978).

Seed size is also related to seedling vigor and stand establishment in wheat. Seedling dry weight 20 days after planting was found to be highly positively correlated with the size of seed among wheat genotypes (Evans and Bhatt 1977). Large seeds have larger endosperm reserves and are thus able to produce larger seedlings than small seeds. The correlation between seed size and seedling vigor among different genotypes is generally not as significant because of genotypic differences in rate of germination and growth (Gul and Allan 1976b). A wheat variety with larger seeds does not, therefore, necessarily have greater seedling vigor than one with smaller seeds.

Research in the Soviet Union indicated that it was possible to increase the drought resistance of wheat by presowing treatments (Henckel 1964). The seeds were soaked in water for 2 days and then air-dried. Plants grown from the "drought-hardened" seed were more drought resistant, with increased protoplasm viscosity, more bound water, higher metabolic rate, and more extensive root systems

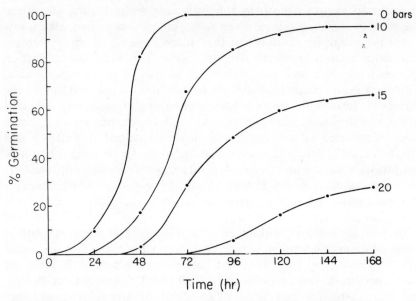

Fig. 55. Germination of "Neepawa" wheat against D-mannitol solutions producing osmotic potentials of 10, 15, and 20 bar and a distilled water control (0 bar).

compared to nonhardened plants. Attempts to confirm these findings have produced variable results. In the United States, little response to presowing treatment was noted, although there were some indications of differing varietal response (Salim and Todd 1968). Another study in Australia indicated that plants from presoaked seeds were slower to develop low relative water contents during drought than untreated plants (Woodruff 1969). The beneficial effects of presoaking may be partially due to faster germination and root growth, which cause faster development of the plant. This in turn permits the plant to escape some of the seasonal drought.

Once the wheat seedling has become established and independent from seed reserves, other factors influence its response to drought. Growth of roots, leaves, stems, and reproductive structures are affected, as will be discussed later. Drought stress can influence the length of the developmental stages. For example, the length of time between floral initiation and anthesis was reduced by moderate water stress, but increased by severe stress (Angus and Moncur 1977). Wheat generally produces more tillers than can ultimately survive and produce spikes. The number of surviving tillers is strongly dependent upon water availability; rate of tiller death increases as water deficit increases (Begg and Turner 1976).

The initiation of floral development in wheat occurs when the seedling has developed about 4 to 5 leaves. Drought stress during the period when the inflorescence is developing can affect certain yield components, particularly number of spikelets. Water stress during the final 1 to 2 weeks before spike emergence can also reduce the number of grains set per spikelet, perhaps not as a direct result of drought but as a result of reduced assimilate availability to the developing inflorescence (Begg and Turner 1976). The net effects of reduced tiller number, which reduces the number of spikes per unit area, and reduced spikelet number depend upon conditions during grain-filling. Some degree of compensation is possible through increased seed size if drought is not too severe.

Root systems of wheat plants, as with many other species, develop almost entirely prior to anthesis. The rate of growth and extent of wheat root systems have been studied extensively, particularly in Canada. In early studies, some 40 years ago, wheat root systems were examined by washing the roots from blocks of soil excavated in the field. The roots of individual Marquis wheat plants penetrated up to 150 cm when grown alone, but only to 120 cm when grown in a crop environment (Pavlychenko 1937). The in situ study of roots in the field by excavation methods tends to be laborious and therefore costly. Many later investigators into root growth grew plants in glass-faced boxes, with the glass front sloped inward so that some of the roots grew along the glass face.

Root studies using glass-faced boxes in artificial environments revealed differences between wheat varieties in their extent and pattern of rooting. The durum variety Pelissier, for example, was found to have an extensive root system, particularly at deeper depths in the profile (Hurd 1964) (Fig. 56). In a comparison of the rooting of common wheat varieties under wet and dry conditions, yield under water-stress conditions was highest in the variety that had proportionately more of its roots in the 20 to 30 cm depth range (Hurd 1968). Root growth of all varieties was reduced by dry soil conditions. Subsequent studies have revealed a broad range in rooting characteristics of wheat varieties. In some cases the yield of particular varieties can be related to root distribution, in other cases yields are not related to root pattern, because of the overriding effects of environmental factors on the aerial portion of the plant.

There has been some controversy as to whether an extensive root system or a more restricted system that conserves water for

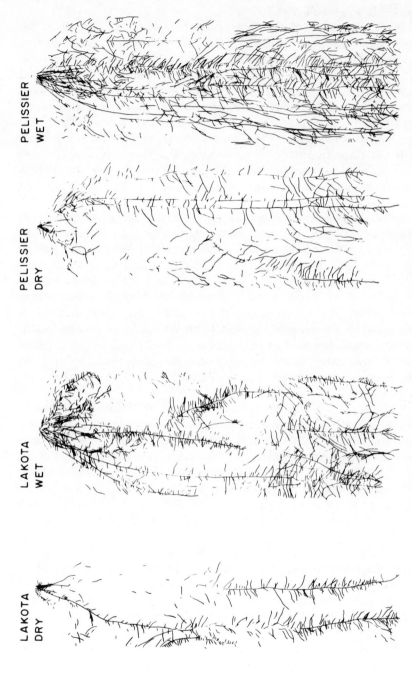

Fig. 56. Tracings of roots, showing on the glass front of root boxes, for cultivars Lakota and Pelissier wheat grown in wet and dry soils (after Hurd et al. 1972).

late-season use is preferable. Results from Canadian studies (Hurd 1964, 1968) have suggested that the extensive root system seems preferable. Studies with wheat in Australia, however, have suggested that restricted root systems may reduce water-use prior to anthesis, leaving more water available during the grain-filling period (Passioura 1972). In this study, root systems of wheat plants grown in a controlled environment were pruned to leave only one seminal root. Seed yield of pruned plants was double that of intact plants. The superiority of either extensive or restricted root systems is largely dependent on the amount and location of stored soil moisture. In many semiarid regions, stored soil moisture exists at depths of up to 2 or more meters. Available soil moisture exists at the 60- to 120-cm depth after wheat harvest in southwestern Saskatchewan (Hurd 1974). Varieties with deep and extensive root systems should be able to exploit this water and provide a yield advantage over less deeply rooted varieties. Indeed, it is widely reported that plant water stress generally decreases as the size of the root system is increased (Bertrand 1965).

As with many other plant species, attempts have been made to determine whether wheat roots are capable of "exploring" for water in dry soil as drought develops. Root growth of wheat reached a maximum between –5 and –10 bar soil water potentials and ceased by –25 bar (Baldry 1973). Other investigations have confirmed that root growth is restricted under severe moisture stress (Hurd 1968). Reports of improved growth of roots as soil dried have appeared in the literature (Hurd 1976). However, soil moisture levels in such studies were near field capacity; drying improved aeration and thus stimulated root growth.

Following the observation that considerable varietal differences can exist in root systems of wheat, research was begun on development of screening techniques to detect these differences more easily. Although studying roots in glass-faced boxes is considerably easier than field excavation techniques, it is still cumbersome when more than a few varieties are to be studied. One approach was to correlate seedling root growth with that at maturity. Such observations indicated that the variety Thatcher produced more roots in the seedling stage than some other varieties, and it also had more roots at maturity (Hurd 1964). Subsequent studies have demonstrated similar results for other cultivars (Hurd 1974). Selection for root-system size is cheaper and faster if seedlings, rather than mature plants, are screened.

While roots are of great importance in supplying water to the aerial portions of the plant, leaves play an important role in the efficiency of use of that water. Morphological and biochemical aspects of leaves, such as control of water loss, surface area, leaf angle, and photosynthetic rate, together determine rate of growth and productivity of the plant. The study of these factors in relation to the yield of wheat under drought conditions has received considerable attention by physiologists.

Much research effort has centered on measuring leaf water status in an effort to determine the basis of water stress in the wheat plant and, subsequently, to determine if genetic differences in drought resistance exist. A simple method of measuring overall leaf water status is to determine the relative water content (RWC). Leaves or leaf discs are sampled from the plant, weighed, floated on water until fully turgid or saturated, reweighed, oven-dried and weighed again. From these weights, RWC is calculated from the formula

$$\frac{\text{fresh weight} - \text{dry weight}}{\text{turgid weight} - \text{dry weight}} \times 100$$

Other indices, known as water saturation deficit or relative saturation deficit can be calculated from the same weights. Where more complex instruments are available, leaf water potential and leaf osmotic potential can be measured. Since the methods of making these measurements have been discussed at length in other reviews (for example, Barrs 1968) they will not be elaborated upon further here.

The leaf water status of wheat plants changes seasonally as drought develops and diurnally as atmospheric demand increases water-loss rate above the rate at which the root system can supply water. Comparisons between leaf water potential of irrigated plants and plants exposed to drought demonstrate seasonal changes in water potential. Water potentials in well-watered plants typically remain in the -10 to -15 bar range, while those of droughted plants drop ultimately to -40 bar or less (Fischer and Maurer 1978). Diurnally, leaf water potential drops as the midday radiation and temperature maxima are approached, and increases again toward sunset and during the night (Millar and Denmead 1976). Intraplant variations in water potential have also been found, with the bottom (older) leaves having lower water potentials than the upper (younger) leaves. If leaf water potential remains low, leaves start to die, or "fire," starting at the base of the plant and progressing toward the top. Under conditions where drought continues to develop through the season, all but the uppermost, or flag, leaf may die. Even the flag leaf may dry up considerably before maturity.

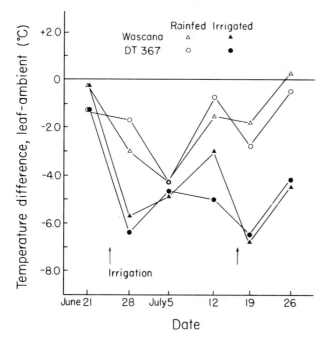

Fig. 57. Leaf and ambient air temperature difference for two durum wheat lines (Wascana and DT 367) measured with a hand-held infrared thermometer.

Leaf temperature can be used as an indirect measurement of leaf water status. As leaf water potential falls below a critical level, the stomata close, restricting water loss. This also leads to a loss of the cooling effect of transpiration, which leads to an increase in leaf temperature. The temperature of well-watered wheat plants is typically a few degrees below ambient air temperature, but rises above ambient in droughted plants. Changes in leaf temperature, as water potential fluctuates, both seasonally and diurnally, can be measured using portable infrared thermometers (Ehrler et al. 1978) (Fig. 57). These leaf-temperature changes can also be used in scheduling irrigation applications, based on measurements made either from the ground or from the air by remote-sensing aircraft or satellites.

Another approach to characterization of stress levels in wheat has been determination of proline and abscisic acid (ABA) contents in stressed and nonstressed plants. Proline levels are considerably higher in stressed than nonstressed plants (Rajagopal et al. 1977), but there seems to be little difference in proline level between different cultivars at the same water potential (Townley-Smith and Hurd 1979). Proline level can serve as an indicator of stress, but, if there are no varietal differences in response, it has little value as a technique for screening for drought-resistant genotypes.

The ABA content of excised wheat leaves increases during wilting. The increase is particularly rapid once leaf water potential falls below -10 bar (Wright 1977). Exogenously supplied ABA will close wheat stomata, and it is thought to play a role in the response of the plant to drought stress (Quarrie and Jones 1977). A comparison of wheat genotypes indicated genotypic differences in ABA content under stress which were related to genotypic differences in leaf water potential (Quarrie and Jones 1979). Also, there was variation among the genotypes in the amount of ABA accumulated under stress. Of 26 genotypes studied, the drought-resistant Canadian variety Wascana accumulated the least ABA, while the German variety Sirius accumulated the most. Since it is easier to screen for leaf water potential than for ABA, it would only be useful to screen for ABA if it proves to be significant as a drought-avoidance characteristic through hormonal control of stomatal aperture.

As leaf water potential falls below some critical level, stomata begin to close. The critical level in wheat varies with leaf age and position, as well as with temperature. Closure occurred at leaf water potentials of -13 to -15 bar in young plants and at -18 to -20 bar in headed plants (Frank et al. 1973). The lowering of the critical level for stomatal closure as the plant matures leads to more rapid depletion of soil water and the early death of leaves (Morgan 1977). There is evidence that the seasonal trend in stomatal aperture, which is commonly reported in terms of leaf conductance, varies between different genotypes (Jones 1977). The seasonal change in stomatal opening and leaf conductance poses a problem to the physiologist: at what time during the season should leaf conductance be measured in order to relate it to plant yield performance? In a study of several wheat varieties, it was observed that yield was negatively correlated with conductance during a 1- to 2-week period prior to anthesis (Jones 1977). This relationship has not yet been demonstrated in other genotypes or in other environments.

Leaf conductance also varies with leaf position on the plant, and between ad- and abaxial leaf surfaces. Leaf position is related to leaf age; lower leaves are older. Stomatal conductance for a leaf is the sum of the conductances of the adaxial and abaxial surfaces, since both surfaces have stomata. In wheat, adaxial conductance is the major component (Jones 1977). Conductance is not related to stomatal frequency, either between varieties or between adaxial and abaxial leaf surfaces.

Leaf shape, orientation, surface texture, and color also influence the water status of plants. Surface hairiness or color affect the albedo of the leaf (Chap. 2) and can reduce radiant heat load and, therefore, transpiration. Upright as opposed to more horizontally inclined leaves will also have a reduced radiation load. Leaf rolling has been observed in drought-resistant durum wheat cultivars (Hurd 1976), a characteristic that reduces transpiration. Reduced total plant transpiration can also be achieved by reducing the vegetative surface area. Wheat generally produces more tillers than it can sustain to maturity, which has led to the suggestion that wheats that are genetically limited to a single culm might be advantageous under drought conditions (Donald 1968).

Investigations of genetic variability in the response of wheat to drought have been made. Differences in leaf water status between particular wheat genotypes can come about from combinations of genetic differences in root systems, efficiency of water transport, stomatal control, and differences in leaf morphology and orientation. Measurements of leaf water potential and osmotic potential, of stomatal diffusive resistance, of leaf temperature, of relative water content (or associated indices), and of water retention by excised leaves have all demonstrated varietal differences in leaf water status of wheat under conditions of moisture stress.

Another area of research on the effect of drought on wheat is the study of the relationship between leaf water status and photosynthesis. In the vegetative stage of wheat plants, the rate of net photosynthesis decreased as leaf water potential dropped below -10 bar, while in the reproductive stage, inhibition was not evident until about -20 bar (Frank et al. 1973). Stomata closed earlier in the vegetative stages than in the reproductive stages. In another study, net photosynthesis of wheat flag leaves declined in a near-linear fashion as leaf water potential decreased; net photosynthesis was near zero at -30 bar potential (Johnson et al. 1974).

Other researches have investigated varietal differences in photosynthetic rate under drought-stress conditions. In an early study, seedlings of wheat varieties differing in drought hardiness did not differ significantly in photosynthetic rate under drought stress (Todd and Webster 1965). Subsequent investigations have demonstrated photosynthetic rate differences related to drought resistance. Integrated net photosynthetic potential of flag leaves of stressed wheat varieties, measured on detached leaves after recovery from the stress, was found to be related to seed yield (Kaul 1974). The measurement of photosynthesis in drought-stressed wheat genotypes by incorporation of $^{14}CO_2$ also demonstrated varietal differences in the ability of wheat to photosynthesize at low water potentials (Dedio et al. 1976) (Table 18). Both of these studies identified high photosynthetic rate under drought stress as a characteristic of the Mexican semidwarf variety Pitic 62. Differences in photosynthetic rate between high- and low-yielding durum wheats during development of diurnal water stress have been demonstrated in the field (Fig. 58).

The relative importance of photosynthesis in leaves, as opposed to other photosynthetic tissues such as the stem and spike structures, varies seasonally and with level of drought stress. Leaf photosynthesis is of necessity the sole source of assimilates in the preelongation growth stage, thus influencing root development, tillering, and early reproductive growth. The rapid senescence of leaves after anthesis, particularly under drought stress, severely reduces the leaf area of the wheat plant. The rate of flag-leaf photosynthesis (Kaul 1974) and photosynthetic area above the flag-leaf node (Simpson 1968) are related to wheat yield. It has often been suggested that preanthesis photosynthate can be stored in vegetative parts of the plant and

TABLE 18: Photosynthetic Capacity of Stressed Wheat Plants as Measured by Incorporation of $^{14}CO_2$ into Leaves

Drought Period (days)	Cultivar	Water Saturation Deficit (%)	$^{14}CO_2$ Incorporation (% of nonstressed Pitic 62)
0	Pitic 62	20.1	100
	Koga	25.1	76
	Neepawa	23.6	69
	Wascana	25.3	64
6	Pitic 62	39.8	60
	Koga	41.6	54
	Neepawa	39.0	51
	Wascana	38.3	43
12	Pitic 62	60.3	32
	Koga	58.7	15
	Neepawa	56.2	18
	Wascana	57.2	17

Source: From Dedio et al. (1976).

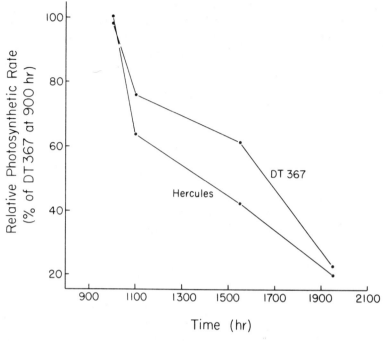

Fig. 58.   Daily variation in photosynthetic rate of a high-yielding (DT 367) and a low-yielding (Hercules) field-grown durum wheat.

then remobilized during grain-filling, and thus contribute to the final yield. This is discussed in more detail in the section on grain-filling.

Having covered vegetative and early reproductive growth, the next major step in the growth of wheat is that of fertilization. Fertilization in wheat, a self-pollinating species, is not as sensitive to drought as fertilization in maize, which is cross-pollinated and has widely separated male and female flowers. The grain-filling period follows fertilization, during which the final economic (seed) yield of the wheat plant or crop is determined. Drought stress can potentially influence yield during this phase by limiting the availability of photosynthate, by reducing translocation, or by shortening the length of the grain-filling period. The period of most active grain-filling in wheat may be only two weeks long, so that reductions in assimilates or the shortening of this phase by only a few days can have a substantial effect on yield.

As pointed out above, both leaf photosynthetic rate and leaf area are reduced during drought stress. Photosynthesis within the spike structures of wheat is less sensitive to drought stress than leaf photosynthesis (Johnson and Moss 1976). It must be remembered that during drought periods, there is still considerable diurnal variation in the water potential of plant parts if water is still available to the roots. Under such circumstances, the plant water potential can be high enough to permit open stomata and photosynthesis during the early morning hours or during cloudy periods of the day. Once all available soil moisture has been utilized, the plant quickly senesces and the seed ripens prematurely.

There has been considerable controversy about the relative contributions to wheat yield of assimilates from current photosynthesis and from remobilized assimilates arising from preanthesis photosynthesis. Estimates of the importance of the contribution of preanthesis photosynthesis to yield, expressed as a percentage of final yield, vary from 3 to nearly 60%. The stem would be the logical site for storage of assimilates until they are required by the developing grain. Circumstantial evidence for retranslocation from stems is provided by experiments showing weight losses of stems of from 30 to 50% between anthesis and maturity and concomitant reductions in concentrations of nonstructural carbohydrates (Austin et al. 1977). These losses are a combination of respirational losses and translocation, and some of the latter probably goes to the grain. Studies of the seasonal trend in nonstructural carbohydrates in wheat stems indicate that they build up to a peak shortly after anthesis, and then decline to very low levels by maturity. The maximum level reached seems to vary with variety, ranging up to 25% of total stem weight. Water stress may influence the relative importance of preanthesis photosynthesis: the preanthesis contribution in one study was found to be 57% in a dry year vs. 35% in a wetter year (Gallagher et al. 1976).

To investigate some of the discrepancies in the understanding of pre- and postanthesis photosynthetic contribution, $^{14}CO_2$ feeding trials throughout the life cycle of wheat plants have been employed. Estimates of the preanthesis contribution based on weight losses in stems and leaves after anthesis ignore respirational losses and physical tissue losses during senescence. A controlled-environment $^{14}CO_2$ study showed that there is some translocation of preanthesis assimilates to the grain, most of which comes from the period between spike emergence and anthesis (Rawson and Hofstra 1969). In field-grown plants, preanthesis contributions amounted to 13% in irrigated and 27% in droughted wheat plants (Bidinger et al. 1977). A higher preanthesis contribution in drought-stressed plants is plausible, since postanthesis photosynthesis is restricted in such circumstances, thus providing more opportunity for remobilization of assimilates from other sources. The aspect of remobilization of preanthesis assimilates and buildup of nonstructural carbohydrate reserves requires more research to obtain unequivocal information on varietal and environmental differences.

The net effect of drought during the grain-filling period is a reduction in yield, arising from lack of assimilates and from premature senescence. Much of the yield reduction results from a reduction in the kernel weight component of yield. In one field study, average kernel weight in drought-stressed plants was 25 mg, compared with 36 mg in irrigated plants (Fischer and Maurer 1978). Reduction in kernel size is greatest in the secondary and tertiary kernels within the spikelet, since they tend to fill later than primary kernels (Sofield et al. 1977).

It is also appropriate to review some of the methodology and aims of physiological studies. The objectives of most drought physiologists working with wheat seem to be twofold: to determine the factors that confer drought resistance, and to develop methods to screen for these factors by the simplest means possible. Simple screening techniques are required for use in plant breeding programs that aim at producing varieties with high yield under drought conditions. The recent availability of improved instruments for measuring plant water status has been of considerable assistance in this regard. Measurements of water potential are still cumbersome, but measurement of stomatal resistance with portable porometers and of leaf temperature with portable infrared thermometers is relatively easy. However, in spite of present technology, it is still only possible to characterize the drought resistance of a few varieties at a time.

One of the problems in studying drought under field conditions has been that of year-to-year and within-year soil variability. This obstacle can be partially overcome by the use of "rain-out" shelters to protect plots from rain. Many researchers have chosen to utilize the almost totally controlled environment of the greenhouse or special growth cabinets. There is growing concern about this because of discrepancies between results obtained in the field and those from controlled environments (Begg and Turner 1976).

A major difference between the field and the controlled environment is the rate at which stresses develop in potted plants in a controlled environment. Withholding water results in a very rapid development of stress, whereas in the field the stress develops slowly during the season. In a study of the grass *Panicum maximum*, potted plants in a controlled environment could reach a stress level in 10 days which would take 4 to 5 weeks to develop in the field (Ludlow and Ng 1976). Various growth response differences between controlled environments and the field have been reported. In a controlled-environment study with wheat, it was concluded that the photosynthetic mechanisms in flag leaves and spikes were affected similarly by water stress (Johnson et al. 1974). However, the same researchers found that under the more gradual development of stress in the field, photosynthesis in the spike was less adversely affected by drought than that of leaves (Johnson and Moss 1976). Under field conditions, the gradual development of the drought allows time for the plant to adapt, during which time root growth and osmotic adjustment can improve drought resistance (Begg and Turner 1976).

Before leaving the topic of wheat physiology, it is perhaps appropriate to briefly take stock of our knowledge of the physiology of drought in wheat and to delineate areas of future research. Studies up to the present have made considerable progress in describing the effects of drought stress on the wheat plant. Some of the major avoidance mechanisms, such as root system size, and tolerance mechanisms, such as photosynthetic sensitivity to water stress, have been identified. There has been continuous improvement in techniques and instruments for studying drought, with the pace of progress responding to each improvement. Knowledge of the effects of drought on biochemical processes in the plant is somewhat more limited. Some progress has been made in the area of photosynthesis, but little on other processes and their associated enzyme systems and control mechanisms.

Perhaps too many of our empirical studies of the effects of water stress on wheat have been done in artificial environments. A survey of the literature reveals that well over half of the drought studies have been done in controlled environments, and in most cases the researchers have made no attempt to confirm their results under field conditions. This can lead to erroneous conclusions, as noted above. This point bears reemphasizing—if physiological studies are to make a practical contribution to improvement of wheat yields under drought, either through breeding or improved agronomic practices, experimental results must relate to what happens under field conditions. In other words, before relying too heavily on them, results obtained in the growth cabinet must be confirmed in the field. This is not to say that controlled environments have no use in drought studies; on the contrary, used with proper caution they can be a useful tool in both research and genotype-screening.

For the future, there is still much research to be done on drought-stress effects. One of the major thrusts should be to continue development of rapid screening procedures which can be used to select wheat genotypes with particular desirable characteristics.

In addition, continued investigation of processes at the biochemical level, such as carbohydrate storage and remobilization, should be made. Readily measured changes in these processes may provide valuable screening techniques for future use. Other drought effects, such as the acceleration of senescence in leaves and other plant parts, warrant further work as well. Physiologists must maintain close liaison with plant breeders for the development of suitable screening procedures and the study of the heritability of the identified drought-resistance characteristics.

Agronomy

Agronomic research with wheat has the objective of maximizing yield under local environmental conditions. This entails optimizing all controllable conditions for the growing crop. Some of these principles apply to any environment—for example, weed control, soil fertility, and rates of seeding—while others are specific to particular environments. This discussion will be limited to the agronomic practices specific to production in drought-prone environments.

Date of seeding is an important determinant of crop growth. In some regions, such as Canada and northern Europe, the choice of seeding dates is quite limited; in other areas with longer growing season or milder climates, seeding can be over a broader time interval. In these latter areas, the date of seeding can be selected on the basis of available soil moisture at seeding and prospects for moisture availability during critical growth stages. In India and East Africa, the data of seeding is determined by the seasonality of precipitation and by the need to avoid the times of highest temperature. Wheat is seeded between October and December in India, after the end of the monsoon rains (Patterson 1965). The crop then matures during the cooler, drier part of the year from November to February. In the highlands of Kenya, rainfall follows a bimodal distribution, with "long rains" from April to August and "short" rains in November and December. Most wheat is seeded at the beginning of the long rain period (Patterson 1965). Australian wheat is seeded during the April–June period and matures before December, thus avoiding the hot midsummer period. In regions with Mediterranean-type climates (winter precipitation) and little or no cold winter weather, wheat is planted from October to December and harvested by June.

In some of the climatic zones with a distinct winter period during which plant growth is not possible, a choice can be made between spring and fall seeding. Many of the wheat-producing areas of North Africa, for example, have a precipitation distribution which is more suited to winter than spring wheat. Winter wheat is seeded in the fall, develops vegetatively before winter, and resumes growth in the spring. It is able to take advantage of early spring moisture to complete development before the onset of summer droughts. In areas where they are adapted, winter wheats generally out-yield spring wheats. The cold-hardiness of winter wheats is a limiting factor in areas with severely cold winters.

Selection of varieties with the appropriate maturity time can be important in some dryland areas. Where the amount of available water is severely limited, early maturity is an important characteristic. In Australia, a study of 12 wheat varieties indicated that from 40 to 90% of the difference in drought resistance between the varieties could be attributed to earliness of maturity (Derera et al. 1969.) Late-maturing varieties exhausted the available moisture before completing their development. Varieties with rapid rates of germination and early growth could be beneficial under such circumstances as well. One disadvantage of many early varieties is that they also tend to be low-yielding.

The actual seeding of the wheat crop requires considerable attention in dryland areas. Important considerations include the necessity of putting the seed into moist soil and at the same time minimizing disturbance of the soil surface to limit moisture loss and reduce chances for soil erosion by wind or water. One of the constraints on wheat yields in semiarid regions associated with manual planting is poor stand establishment. Mechanized planting (Fig. 59) can alleviate many of the problems associated with manual planting, such as uneven seed spacing, uneven planting depth, and poor seed covering and packing.

In situations where the soil surface is dry at planting, it is desirable where possible to place the seeds in moist soil beneath the dry layer. The maximum seeding depth is dependent in part upon varietal differences, such as coleoptile length (Allen et al. 1962). Coleoptile length and plant height are correlated, so semidwarf wheat varieties cannot be planted as deeply as standard varieties. Winter wheat is often planted as deeply as 10 to 13 cm in the northwestern United States. The seed is placed in the bottom of deep furrows created by specialized furrow drills.

In many semiarid regions, interest is growing in the use of minimum-tillage systems for growing wheat. This involves the use of the minimum possible number of tillage operations, often associated with the use of chemicals for weed control. Minimum-tillage reduces the production costs through reduced fuel use, conserves soil moisture

Fig. 59. Traditional and improved seeding methods in India. The bullock-drawn seeder permits even metering of seed and fertilizer with placement at a uniform depth in moist soil.

through less disturbance of the soil surface, and helps to reduce erosion since crop residues are left on the soil surface. In southwestern Saskatchewan, yields of wheat were not reduced by eliminating preseeding tillage (Anderson 1975).

The rate of seeding and plant geometry within the crop stand have received research attention in relation to drought conditions. Globally, seeding rates of wheat vary from 17 to 200 kg ha^{-1}, the lower rates being common in dry areas, the higher ones in moist areas (Patterson 1965). In a comparison of seeding rates of 22, 45, 67, and 101 kg ha^{-1} in southern Saskatchewan, yields of the lower two seeding rates were higher than those of the upper two (Pelton 1969). The rate of moisture use was lower at the low seeding rates than at the high seeding rate, leaving more soil water for development and maturation of the seed. Low seeding rates generally result in later maturity, which can be a disadvantage in areas with a limited growing season. Adequate weed control is also required at low seeding rates, since the crop is less able to inhibit weed growth through competition.

Plant spacing and row orientation can also influence wheat growth and yield. Increased row width at constant spacing increases within-row plant competition. The optimum row spacing/seeding rate combination varies with environment. Under dry environments, both excessively low and high plant densities reduce wheat yield (Puckridge and Donald 1967). The direction of planting can affect leaf water status and yield of wheat (Erickson et al. 1979). Rows planted in a north–south direction had higher leaf water potentials and lower stomatal resistances than other row directions. Grain yield, however, was maximum in east–west rows.

Soil fertility level and adequate weed control in the growing crop influence water-use efficiency and yield. Water-use efficiency can be improved by applying fertilizer to nutrient-deficient soils where the crop yield is more reduced by the nutrient deficiency than by drought stress. Concern has been expressed that the use of excessively high rates of nitrogenous fertilizers may induce a stimulation of vegetative growth which would in turn increase water use and lower crop yield. In Australia, reduced levels of fertilizer N and P application reduced evapotranspiration but caused a relatively greater reduction in plant dry-matter production (Fischer and Kohn 1966a). Yield was reduced in some cases by high N and P applications (Fischer and Kohn 1966b). The application of high rates of fertilizers to dryland wheat crops is likely to have significant adverse economic consequences long before impairment of crop growth occurs.

Weeds will reduce yields of dryland wheat crops by competing for water, nutrients, and growing space. This competition occurs both below and above the soil surface. The total length of roots of wheat plants competing with wild mustard was found to be less than 50% of that of plants growing without the weed competition (Pavlychenko 1937).

Water management is a very important aspect of agronomic practice in relation to dryland wheat production. The primary aim in water management is to make the most effective use of available water for crop production. This can involve fallowing and/or snow management to improve soil water storage, and evaporation control by such means as wind barriers.

Fallow refers to keeping the land bare of vegetation for a crop season in order to conserve soil moisture for production of a crop in the following season. Water, from precipitation, which falls on the bare soil during fallow may either be conserved in the root zone, percolate below the root zone, evaporate from the soil surface, or be used by weeds. Water can be lost as surface runoff or as snow blowing in cold climates (Staple 1960). For fallow to be economically effective, crops grown on fallow must yield at least twice as much as crops grown continuously.

The amount of water stored during the fallow period depends on the amount and type of precipitation, the presence of crop residues on the soil surface, and the soil texture, among other things. In southern Saskatchewan, the amount of precipitation conserved during the normal 21-month fallow period (from harvest of the previous crop to planting of the next) averages 21% of the total, or some 100 mm (Staple 1960). More than 50% of the storage takes place during the first fall and winter, and relatively little during the final winter. Similar storage patterns have been found in the northern United States. In Australia, moisture storage during fallow periods ranged from 1 to 19% of precipitation. Storage was low in coarse-textured soils and higher in fine-textured soils (French 1978).

The significance of the amount of water conserved by fallowing is determined by the yield of the ensuing wheat crop. Research in southern Saskatchewan indicates that about 140 mm of water is required to produce the minimum amount of grain in wheat, and each additional 25 mm produces about 270 kg ha^{-1} of grain (Staple and Lehane 1954). In drier regions with higher evaporative demand, the minimum requirements for producing the crop are often greater and the increments from added water are smaller. In many instances, the water stored during fallow means the difference between acceptable and unacceptable crop yields. Under conditions of more favorable moisture the drawbacks of fallowing, such as accelerated loss of organic matter, increased soil salinity, and cost of tillage, outweigh its benefits.

Moisture storage in fallow can be improved by management practices that reduce losses due to runoff, snow blowing, and, to some extent, evaporation. Runoff during the fallow period can be reduced by contour tillage or by land-shaping. Land-shaping, involving the construction of level benches fed by runoff from an adjacent slope, is expensive and has been more successful with perennial forage crops than with wheat (Haas and Willis 1968).

In North America and the Soviet Union, snow management in the dryland wheat-growing regions can provide more water for crop production. This basically involves trapping blowing snow. Snow is trapped by the use of plant barriers placed at right angles to the prevailing wind direction (Fig. 60) or is prevented from blowing by maintaining standing crop stubble. Plant barriers range from permanent types, consisting of rows of trees, shrubs, or tall perennial grasses, to temporary barriers made by alternating the stubble height during harvest or by leaving narrow strips of the crop standing. The amount of snow trapped by barriers is a function of their height, density (or porosity), and spacing.

Stubble height left after harvest of the wheat crop influences the amount of snow held and thus the amount of water stored. In North Dakota, water storage was increased by 28 and 43 mm by 25 and 50 cm stubble heights, respectively, above that of bare soil (Willis et al. 1969). The high moisture storage during the first winter of the fallow period in Canada is due in part to the presence of stubble to trap snow (Staple 1960). Retention of a stubble mulch on the soil surface through the fallow period also helps to limit surface evaporation and prevent wind erosion. The use of herbicides, rather than tillage, to control weed growth during the fallow period helps to conserve the crop residues (Anderson 1971). The economics of this practice depend upon the relative costs of the herbicides and of the equipment and fuel for tillage.

The use of windbreaks to alter crop microclimate has been investigated for wheat, as well as for numerous other crops. In a pilot study, slatted fencing reduced windspeed by 15 to 49% and evaporation by 12 to 23%, at the same time increasing wheat yields by 24 to 43% (Pelton 1967). Leaf water potentials were found to be lower in exposed than in sheltered wheat plants. Windbreaks that reduced wind speed most significantly produced the highest leaf water potentials in the wheat canopy (Frank and Willis 1972). Similarly, in winter wheat, windbreaks improved leaf water status,

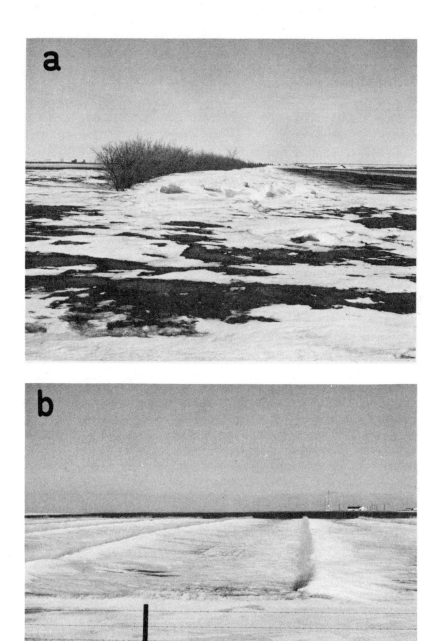

Fig. 60. Wind barriers for trapping snow. (a) Perennial shrub barrier. (b) Barriers formed by a perennial grass (*Agropyron elongatum*).

but there was no consistent yield increase noted in sheltered as compared to exposed plants (Skidmore et al. 1974). Windbreaks are also valuable in limiting wind erosion in light soil and in trapping snow in colder climates. Considerations in the planning of windbreaks include selection of tree or shrub species with the appropriate height, spacing of the plants within the windbreak, and spacing of the windbreak rows. The porosity of windbreaks should be 40 to 50%; excessively dense windbreaks have poorer air-flow characteristics and waste land (Sturrock 1975). If the only reason for installation of windbreaks is to increase yield, the expected yield increase must be weighed against the cost of installation and the amount of land used by the windbreak.

The future mandate of wheat agronomists will, in some respects, remain the same as it has in the past: to optimize wheat yields within climatic and resource limitations. Energy is one important limiting resource which is becoming critical, and which will become more critical in the future. At the present time, the rapidly accelerating cost of fossil fuels is the major consideration, while in the future both availability and price will be important. The highly mechanized production of dryland wheat in the developed countries consumes large amounts of energy; calories recovered may be less than input. The transportation of grain to world markets also consumes large amounts of energy. Consumption of fossil fuels is relatively less in developing countries, but costs still increase prices of fertilizers and of fuel for transportation systems. The energy problem gives a new challenge to the agronomist to develop production techniques which will reduce energy requirements without reducing yields.

Breeding

In general, the objective of wheat breeders has been to improve yield. It must be added that in some countries, particularly Canada, this is yield improvement within fairly rigidly controlled quality classes of wheat. The quality classes vary with intended use and with differences in processing technology and consumer preference. Increases in the yield of wheat by genetic improvement can arise from many factors such as insect and disease resistance, improved photosynthetic capability, improved partitioning of photosynthate between vegetative and reproductive growth, and improved resistance to environmental stresses such as drought. We will first examine some of the approaches used in breeding drought-resistant wheats and then assess progress and future prospects.

Breeding for drought resistance in wheat means breeding for yield in environments where water deficit is the major factor limiting production. Improvements in yield come from recombination of genes by crossing different genotypes, followed by selection of the progeny possessing the desired agronomic features and a higher yield level. Due to the large numbers of plants involved and the complex interaction of characteristics associated with drought resistance which make selection for drought resistance difficult, the breeder is essentially limited to selecting for yield, which is relatively easy to measure. Yield reflects the integrated effects of the many factors that contribute to it, including root system size, stomatal control of transpiration, photosynthetic efficiency, and translocation into the seed. It is very difficult to choose any single factor that contributes to high yield and select for it in a breeding program (Boyer and McPherson 1975). Optimizing this one characteristic may do little good since other characteristics become limiting. Breeding for drought resistance is somewhat environment-specific as well. The most drought-resistant variety in a particular environment may not be the most resistant in another due to differences in the time of drought in relation to growth stage. Progress in breeding for drought resistance is likely to be most rapid in areas where the drought occurs at approximately the same growth stage year after year (Boyer and McPherson 1975).

A breeding program aimed at improving drought resistance of wheats was initiated at Swift Current, Saskatchewan, some 10 years ago by E. A. Hurd. Hurd took the approach of carefully selecting parents on the basis of characteristics conferring adaptation to drought, making the crosses, and then growing and handling large populations for yield testing under dryland conditions (Hurd 1971). Selection for yield was started as early as the F_3 generation.

Careful selection of the parental genotypes to be used in a crossing program can increase the chance of finding superior progeny (Wallace et al. 1972). If parents that have complementary physiological characters for drought resistance are used, the chances of obtaining high-yielding progenies should be improved. Due to the vast numbers of different individuals in the segregating populations arising from a cross, it is not possible to select for any physiological characteristic at this stage. This is primarily because easily measured characteristics of drought-resistant genotypes have not been found.

Physiologists, then, can make their greatest contribution to breeding for drought resistance by identifying readily measurable characteristics which contribute to adaptation to drought. Once this is done, potential parents can be screened for these traits and the breeder can use the most promising ones in crossing programs. Although this may sound simple enough, in practice it is not. Aside from the problems of creating a realistic environment for studying physiological parameters, such studies require a considerable investment of manpower—a relatively scarce commodity at most research centers. Technological advances in instruments for measuring parameters such as leaf water status are of some benefit in reducing the labor requirement. The development of rapid and simple screening techniques remains a high priority.

Once the cross has been made, it is necessary to handle large populations in the segregating generations to increase the chances of finding desirable lines. While the F_3 generation may be the preferred time to commence yield testing, this is often not possible in semiarid climates since the F_2 plants do not produce enough seed for replicated yield (Hurd 1976). Under Saskatchewan conditions this is generally true, so the F_3 generation is increased during the winter in California or Mexico and yield testing commences with the F_4. The yield testing of large populations involves a considerable manpower commitment, but it is certainly much less than would be required to select for drought-resistance characteristics in segregating populations.

Selection for drought resistance in segregating populations may be possible in certain situations. Large numbers of lines could be screened for ability to germinate against osmotic stresses, or for root growth in the early seedling stage (Townley-Smith and Hurd 1979). As yet, methods such as this have not produced drought-resistant varieties of wheat.

Can we select for drought-resistant varieties that will have superior yield performance under optimal growing conditions? In other words, can we have varieties that will yield well under dry or moist environments? Experience indicates that selecting for high yield under optimal growing conditions may not necessarily find factors that will also improve yield under dry conditions. In one study it was reported that the highest-yielding wheat line under adequate moisture was the lowest-yielding under drought stress (Hurd 1969). A reaction such as this indicates a strong genotype–environment interaction. Variation in stability of yield of wheat varieties grown in different environments has been reported (Johnson et al. 1968). Comparison of a number of wheat varieties under wet and dry environments at a single location revealed that some were capable of a high yield ranking under both environments, while others exhibited marked change in rank from one environment to the other (Table 19). A variety with a stable high yield across environments would be very desirable, particularly

TABLE 19: Yields and Yield Rankings of *Triticum aestivum* and *Triticum turgidum* Cultivars under Rainfed and Irrigated Conditions at Swift Current, Canada

	Rainfed		Irrigated	
	Yield (g)	Rank	Yield (g)	Rank
T. aestivum				
NB 320	581	1	1,284	2
Canuck	502	2	884	9
Pitic 62	491	3	1,339	1
Echo	490	4	995	4
Amy	476	5	984	5
Neepawa	430	6	906	8
Sinton	419	7	853	10
Koga	402	8	944	6
ACEF-125	388	9	1,020	3
Glenlea	318	10	917	7
T. turgidum				
DT 363	567	1	1,290	1
Cando	528	2	1,107	3
DT 367	505	3	1,153	2
Pelissier	498	4	891	8
Wascana	436	5	1,036	4
Wakooma	404	6	964	5
Lakota	396	7	952	6
Hercules	387	8	919	7

where droughts occur sporadically from year to year. However, the probability of finding such a variety is low (Hurd 1976).

Another suggested approach to breeding of wheat and other crops is the development of a model plant or "ideotype" (Donald 1968), as opposed to breeding to eliminate defects or to improve yield. This involves the combination of all the characteristics deemed desirable under a particular environment into a single genotype. To develop an ideotype concept for a crop requires a great deal of background physiological information. If this can be accomplished, the breeder is then faced with determining the genetic variability for the desired characteristics, assessing their heritabilities, and then combining them into a single genotype.

An ideotype for dryland wheat in Australia has a short, strong stem; few, small, erect leaves; a large, erect spike with awns; and a single culm (Donald 1968). At the present time there is sufficient physiological information to confirm the desirability of some of these features, but not others. Possession of a few small leaves, for example, is not likely to inhibit yield, since the current consensus is that yield of wheat is not being limited by assimilative capacity. The proposed reduction in leaf size goes against the trends in the evolutionary development of wheat and its relatives: leaf size has increased with the development from primitive to modern types, but rate of photosynthesis per unit leaf area has fallen (Evans and Dunstone 1970). Wheats that produce a single culm could be of benefit, but only in fairly stable environments where factors such as available water are reasonably constant from year to year. Under conditions where drought stress increases as the season progresses, uniculm types could have an advantage. Normal plant types produce several tillers early in the season, but, as the drought increases, many of these die and do not contribute to yield. Elimination of such wasteful tillering could help conserve water for use later in the season.

The ideotype concept has been in use in a limited form for some time. In fact, the drought-resistance breeding method suggested above involves the ideotype concept: parents with desirable and complementary characteristics are chosen to improve the chances of finding progeny with the desired attributes. The development of a complex ideotype is a very long-term proposition because of the problems involved in breeding. Indeed, we cannot even propose a detailed ideotype at this time because of our incomplete knowledge of the physiology of the wheat plant.

Having outlined some of the approaches to breeding drought-resistant wheats, let us now turn to what has been accomplished in this area. If it is decided that the development of any wheat variety that has improved yield under drought conditions constitutes progress in breeding for drought resistance, many examples of such varieties could be cited. However, in many such cases, the breeding programs were probably conducted without very much direct consideration of drought resistance in either the parental material or the progeny. In other cases, the breeding program has been directed at improving drought resistance, and a few examples of the successes can be given. Progress has been relatively simple in the drought-avoidance characteristic of earliness of maturity, but somewhat more difficult in the case of root system size.

An example of the development of a variety with improved drought resistance as a result of improved root system size is the durum variety Wascana, bred in Saskatchewan. This variety is a

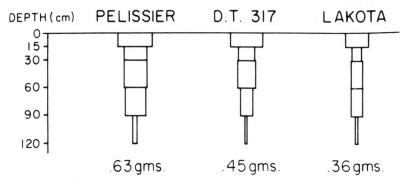

Fig. 61. Weight of roots of mature plants of Lakota, Pelissier, and a Lakota X Pelissier offspring (DT 317) grown in a dry soil. Bar width represents the amount of root at each level in the soil (dry wt/plant).

selection from the cross Lakota X Pelissier (Hurd et al. 1972). Pelissier had previously been identified for its extensive and deep root system (Hurd 1964). Comparisons were also made of the root patterns of Lakota and Pelissier, and of Lakota X Pelissier crosses (Hurd et al. 1972). Pelissier was found to have a consistently larger root system than Lakota, and the Lakota X Pelissier lines had root systems more like Pelissier than Lakota (Fig. 61). The yield of Wascana is similar to Pelissier under severe drought, and superior under less severe conditions. This variety resulted from the careful choice of parents, followed by selection for yield under dry environments.

The future pace of progress in breeding for drought resistance in wheat will be constrained by the same factors that were present in the past: physiological knowledge, availability of manpower, and restrictions on crop quality. Advancement in physiological knowledge is limited by many factors. The major factor is probably lack of both professional and technical manpower. Plant breeding progress is similarly constrained. The present trend in many Western countries is reduced or static support for research, with no immediate change foreseeable. This trend may be reversed in the future as concern over global food supply grows. Increase in wheat yields in Canada is also severely constrained by the restrictions on quality within the various wheat classes. Some very high-yielding wheat genotypes have been developed, but since they do not have acceptable quality, they have been discarded. These quality restrictions are unlikely to be relaxed until the global food situation becomes more critical.

Summary

In looking at trends in wheat yields over the past several decades it is difficult to ascertain what proportion of the yield improvements can be attributed to breeding of improved varieties and what proportion to improvements in agronomic practices. The proportions are also likely to be quite different between locations, which in turn are related to previous breeding and agronomic research activity.

An analysis of wheat yields at Swift Current, Saskatchewan, from 1922 to 1972 showed that most of the variations in yield could be attributed to weather trends (Fig. 62), leaving little which could be accounted for by technological advance (Robertson 1974). Nevertheless, there has been some yield increase, part of which is due to modest increases in yield potential and the remainder to agronomic factors such as fertilizer application and weed control. The genetic advance in yield of Canadian hard red spring wheats was 15% between 1908 and 1969. Yield of wheat increased by 80% in the United Kingdom between 1948 and 1978 (Austin 1978); between 1908 and 1978, 40% of the improvement in yield could be attributed to genetic improvement alone (Austin et al. 1980). In the state of New York, wheat yields doubled between 1935 and 1975 (Jensen 1978). This increase was attributed almost equally to technological improvement and genetic advance. The average wheat yield increase in the United States as a whole has averaged about 36 kg ha^{-1} yr^{-1} over the past several decades.

Improvements due to technological advance are more evident in parts of the developing world, where even seemingly minor changes in agronomic practices coupled with improved varieties have greatly increased yields. Such rapid improvements are to be expected where the original yield levels were poor. In Mali, for example, traditional wheat varieties yield about 1.5 tonnes ha^{-1}, whereas improved varieties, using traditional cultural practices, yield 3.5 to 5.0 tonnes ha^{-1} (CIMMYT 1977).

The greatest achievements in developing drought-resistant wheats thus far have come from research institutions where physiologists and breeders have been working as a team. Such teamwork is the only hope for rapid progress in the future. Physiologists working in isolation from breeders are generally not aware of the complexities of plant breeding. The former tend to concentrate on empirical studies of the effects of drought on plants, usually under artificial environmental conditions, and do not consider genotypic differences in response to drought. Similarly, the breeder working alone often does not have the time or expertise to conduct drought physiology

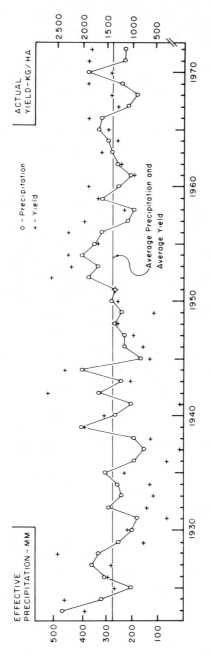

Fig. 62. Annual yield of wheat on summerfallow at Swift Current, Saskatchewan, in relation to annual effective precipitation (after Robertson 1974).

studies. The team approach ensures that the physiological studies are relevant to the natural environment and of use in screening parental material for breeding programs. Agronomic input is required when new varieties are produced to ensure that optimal management techniques are used with the new variety.

SORGHUM

Introduction

Sorghum (*Sorghum bicolor* Moench) is currently ranked fourth among the world's major cereal crops. The 3-year (1977 to 1979) global average production was 68.7 million tonnes (Hulse et al. 1980). Increases in acreage and productivity have been spectacular in the United States and Argentina due to the commercial exploitation of hybrid vigor, following the discovery of cytoplasm-gene-regulated male sterility in the 1950s (Rao and House 1972). Aside from the use as feed grain in these two countries, sorghum is the staple diet for about 450 million people in the semiarid tropical regions of a number of the poorest developing countries. While sorghum probably had its origins as a domesticated crop in Abyssinia (Doggett 1970), it is now an important crop throughout Africa, much of India, China, the Middle East, and more recently Central and South America.

Two principal reasons for the importance of sorghum in the semiarid tropics are the drought-resistant properties of the species and the diversity of forms which can provide grain, forage, sugar, or fuel. Sorghum is of great nutritional significance to that diverse group of people who occupy the semiarid margins of the desert areas of the globe where drought is a frequent event. Interest in understanding the nature of the drought resistance in this important crop has increased sharply in the last 10 years, as plant breeding and the spread of hybrids have improved the genetic threshold of yield. At the same time concerted international efforts, such as the formation of the International Crops Research Institute for the Semi-Arid Tropics at Hyderabad, India, have been directed toward improving sorghum in the interests of the many poor people in developing countries who are highly dependent on it for their nutrition. The aim of this section is to give some indication of the nature and extent of the current research into the drought-resistant properties of this species.

General Effects of Drought on Growth and Yield of Sorghum

Most of the general approaches used to investigate the drought-resistant qualities of sorghum have depended primarily on field experiments where, for example, sorghum may be contrasted with other field crops in a variety of rainfed situations by means of growth and yield analysis. Alternatively, rainfed sorghum may be contrasted with a sorghum crop given various levels of irrigation. Both these latter approaches may require a series of observations in successive years until a growing season with a pronounced drought is encountered. Drought can be simulated under field conditions by the use of rain-out shelters (Arkin et al. 1976). Detailed comparisons of phenology and yield changes caused by drought stress can be made in controlled environment conditions where a range of specific water deficits can be tested when light, temperature, and nutrient status are systematically varied. A significant advantage of a controlled environment is that the conditions can be fairly accurately reproduced on successive occasions. The disadvantages are that it is difficult to simulate all the natural characteristics of a field situation where large bodies of variable soil, crop competition, and continuous variation in radiation inputs, wind speed, and relative humidity create complex interactions.

Evidence from these different approaches indicates that, as with other crops, water deficits reduce the potential yield of sorghum. The magnitude of the reduction varies according to the degree of interaction between the drought stress and such variables as stage of life cycle, planting density, genotype, and previous adaptation to drought. In general it can be said that sorghum belongs to a class of crop plants that can survive fairly severe water deficits and still produce a modest yield. Some examples of studies aimed at measuring the effects of drought stress on the general characteristics of the plant, such as yield, are given in the remainder of this section as an introduction to the more specific and detailed approaches dealt with in later sections.

The following is an example of a field study (Eck and Musick 1979a) in which plants stressed under low rainfall conditions were contrasted with irrigated plants. The results indicated that yield reductions of grain differ in magnitude according to the stage of the life cycle affected by water stress. When plants were given moderate stress periods of 13 to 15 days in which the average afternoon leaf water potentials (ψ_w) were between -15.8 and -20.3 bar, there was no reduction in grain yield. However, if the periods of stress were extended to 28 days with an average leaf ψ_w of about -21.7 bar, beginning at the early boot or heading stage, yields were reduced by 27%. The same length of period and level of stress beginning at early grain-filling only reduced yields by 12%. Longer periods of more severe stress (35 days at -22.9 bar, or 42 days at -24.0 bar) beginning at the boot stage, reduced yield markedly by 43 and 54%, respectively. The authors noted that yield reductions from stress initiated at the early boot stage were the result of reduced seed size and seed numbers. Stress at heading or later only reduced seed size. They also concluded (Eck and Musick 1979b) that reduction in grain yield by stress was associated with a decrease in accumulation of nutrients into leaves.

Although the general growth of sorghum can be reduced by water stress, the effect on grain yield can be partly buffered by the ability of the plant to adapt to stress by diverting dry matter into the head even under conditions of considerable stress. Fereres et al. (1976) indicated that under conditions where water stress built up slowly to -20 bar leaf ψ_w, plants adapted so that stomates still remained open, despite the fact that at earlier stages in the life cycle stomata closed at -16 bar when stress was developed rapidly. Under the conditions of slow development of stress, grain yield differed by only 10% between irrigated and stressed plants, whereas the total differences in dry-matter production were about 26%. The same authors (Fereres et al. 1978) indicated that sorghum, as contrasted with maize, can maintain cell turgor even at grain-filling because of a superior ability to adjust solute potential (ψ_s) throughout the season as a physiological response to increasing drought stress. Maintenance of cell turgor by osmotic adjustment in the face of increasing water stress permits stomata to remain open, even at -20 bar leaf ψ_w, so that photosynthesis can continue.

Several field studies by Inuyama, who investigated the effects of water stress on grain yield, indicate different responses by different genotypes of sorghum, also different responses to stress when sorghum plants are grown at different densities. A comparison of varieties (Inuyama 1978a) under similar water-stress conditions indicated differing abilities to yield grain and different degrees of response to stress. Inuyama attributed the yield differences to the higher leaf diffusive resistance (R_L) (at leaf ψ_w below -19 bar) of the higher grain yielder, which prevented water loss during the drought-stress period. This effect was most marked at the sensitive boot stage. A resistant variety was not only less affected by a given stress treatment than a less resistant variety, but also was less affected by water stress if exposed at any given growth stage. Inuyama et al. (1976) demonstrated that a critical period of growth, affected by drought stress, exists at the boot stage and continues through blooming. The principal effect of drought was to decrease the number of seeds per head; sorghum appeared to have a limited ability to compensate for reduced head size by increasing grain weight despite water stress. In another field experiment where the effects of water stress were examined in plants grown at different densities, Inuyama (1978b) showed that plants from high-density plots, when given drought stress, had lower leaf ψ_w and higher R_L and suffered more severe stress effects than plants from low-density plots. Grain yields were highest in low planting-density plots for both stressed and nonstressed conditions. Again the negative effects of drought stress were greatest when stress occurred at the boot stage.

Field studies at ICRISAT, India (Sivakumar et al. 1979), comparing sorghum grown under irrigation or on post-rainy season residual soil moisture also indicate that water stress considerably reduces yield. Plants in irrigated plots had higher stomatal conductance and higher leaf ψ_w than plants grown on limited soil water. Stressed plants had lower yields, less total dry matter, and a reduced leaf-area index compared with irrigated plants. Water-use efficiency was decreased by continual exposure to water stress: on a seasonal basis nonirrigated plants used 213 mm of water to produce 0.51 kg m^{-2} of dry matter, whereas irrigated plants extracted 321 mm of water to produce 0.93 kg m^{-2} of dry matter.

Similar investigations by pot experiments (Langlet 1973) indicated that the most sensitive period for drought stress to affect grain yield occurred during the grain swelling to dough stage. Earlier stress periods longer than 20 days duration reduced the number of grains and the kernel weight.

Growth analysis comparisons of hybrids with parents (Gibson and Schertz 1977) indicate that hybrids develop faster and produce greater total dry matter per season than parent lines. This suggests the probability of differing responses to drought stress. This is borne out by a study, using rain-out shelters to induce drought stress on 5 sites, which showed that hybrids had a wider adaptation to drought stress than lines (Jika et al. 1980). The ability to adapt to prolonged drought stress in several ways permits sorghum to reach a relatively high yield under conditions where crops such as soybean and maize may show marked reductions in yield (Hatfield et al.

1978); the ability to cool leaf temperatures below ambient air temperature is one of these adaptations. In this study, when soybean, maize, and sorghum were compared under field conditions, canopy temperature differences were noted between the three species under both irrigated and water-stress conditions.

The above approaches to understanding drought-stress effects in sorghum indicate that response to drought stress can vary according to genotype, stage of development, duration and severity of drought, degree of previous adaptation to drought stress, planting density, temperature, and many other factors. To understand the nature of these responses to drought stress it is necessary to examine more closely the effects of drought on specific stages of the life cycle. This is the subject of the next four sections.

Germination

Evidence from studies in soils with varying degrees of moisture and from germination studies in solutions with a range of osmotic potentials indicates that there are significant cultivar differences in the ability of sorghum seeds to germinate in the presence of soil water deficits. Mali et al. (1978) compared 14 cultivars in soils with a range of water potentials and found there was a specific hydration level for each cultivar below which germination would not occur. As the water potential of the soil was lowered, germination percentage decreased in all cultivars. The critical seed moisture content below which germination did not occur ranged between 20 to 30%. In a similar type of experiment, Stout et al. (1980) showed that at 20% soil moisture content germination was normal for two cultivars, but at 17% moisture content one cultivar failed to germinate and the other was reduced to 60% germination.

A number of studies have used osmotic solutions to lower water potential to simulate soil dried to various levels of moisture content. These kinds of studies also indicate significant cultivar differences in germination behavior. Wiggans and Gardner (1959) and Stout et al. (1980) demonstrated cultivar differences in both germination rate and percentage germination as water potential was decreased with various osmotica. Stout et al. (1980) showed that the osmotic solutions had little effect on the total amount of water uptake by the seed, suggesting that part of the germination inhibition was due to secondary effects of the osmoticum rather than a simple limitation on water availability. El-Sharkawi and Springuel (1977) demonstrated that when polyethylene glycol is used to simulate reduced matric potential, temperature has an important interaction such that the lowest water potential at which seeds can germinate depends on temperature. They noted that optimal temperatures for permitting germination at reduced water potential were higher for sorghum than for wheat. Plumule emergence was more sensitive than the radical to reduced water potential. The threshold for plumule emergence in sorghum (-10 bar) was lower than for wheat (-7 bar), suggesting that sorghum can germinate at lower soil water potentials than wheat.

There is some information to indicate a significant relationship between the seed germination characteristics under soil moisture stress and the subsequent response of the seedling or adult plant to drought stress. Saint-Clair (1976) found that the range of response of 11 sorghum cultivars germinated at different water potentials in polyethylene glycol solutions correlated with a range of field drought-tolerance characteristics of the same cultivars. Stout et al. (1980) and Saint-Clair (1976) both demonstrated that the cultivar NK 300, which has the characteristics of a fast-growing drought avoider, had better germination than the slower growing, more drought-tolerant cultivar M-35 under similar conditions of osmotic stress.

It can be concluded then that germination of sorghum seed is sensitive to soil moisture stress and that some cultivars are better adapted to this stress than others.

Roots

It is generally assumed that increased root growth is an important mechanism for drought avoidance. Lateral movement of water in soil is slow. By increasing lateral root density and length, per unit volume of soil, the plant can recover more soil moisture and offset increased transpirational loss due to increasing evapotranspiration potential. In addition it is claimed that reduced resistance to water flux through the roots can be achieved in plants by a lowered radial resistance (Williams 1976).

Sorghum is known to be a more drought-resistant crop than soybean, and Teare et al. (1973) have shown that sorghum plants have approximately twice the weight of roots per unit volume of soil as soybean plants. They noted that the water-use efficiency of sorghum (gm dry matter/kg H_2O) was approximately 3 times that of soybean on a dry-matter or grain-yield basis. In these particular experiments the leaf diffusive resistance of sorghum was almost 3 times greater than soybean, indicating considerable control of water loss from leaves despite the larger root volume.

Other work has indicated the importance of root growth in drought resistance of sorghum. Nour and Weibel (1978) demonstrated that drought-resistant types of sorghum had heavier root weight, greater root volume, and higher root:shoot ratios than the less resistant cultivars. Bhan et al. (1973) compared 8 cultivars drought-stressed in field conditions and demonstrated that the most drought-resistant cultivars had deep, penetrating roots and a greater total root weight and more extensive primary and secondary root systems than the less drought-resistant cultivars. In a somewhat similar kind of comparison, Saint-Clair (1977) found that cultivars widely adapted to a variety of sites in a semiarid or dry environment had a good balance between shoot and root growth (as estimated by root:shoot ratios determined on a dry-weight basis). On average, about 84% of the root weight was found in the top 25 cm of soil, indicating a high degree of reliance on the upper profile of the soil for moisture.

Seminal roots seem to be significant only in the very early seedling stage of growth. For example, even under nonstressed conditions the seminal roots cannot support normal growth for very long after seedling establishment, and adventitious root growth is essential for normal growth (Bur et al. 1977; Chotib et al. 1976). Changes in the distribution of the adventitious roots can occur in response to the changing soil moisture profile. For example, in a comparison of frequently irrigated versus sparingly irrigated sorghum, the root-length density was actually greater in the upper soil profile of the less irrigated treatments than the frequently irrigated plants.

Field experiments to compare genotype differences in root-growth response to drought stress are difficult to carry out due to the practical difficulty of recovering root material from the full soil profile. Sullivan and Ross (1979) have devised a hydroponics system where water stress can be imposed by the use of polyethylene glycol solutions in a manner that achieves a close correlation between plant responses in the controlled environment and field-grown plants. The plants are grown in tall, slender light-proof polyvinyl chloride tubes filled with aerated nutrient solution. Water availability is controlled by the inclusion of various quantities of polyethylene glycol (Carbowax 600) as an osmoticum in the nutrient solution. Using this system it is possible to quickly evaluate main root numbers, secondary branching, maximum root length, root volume, dry weight, and root:shoot ratios. Sullivan and Ross (1979) used this system to compare the growth characteristics of two genotypes considered to be relatively drought resistant. They found that the root characteristics were particularly significant in relation to drought avoidance, rather than to tolerance. They found significant differences between the cultivars with respect to the ratio of shoot:root length and shoot:root dry weight which were consistent with the hypothesis that increased root growth and thus drought avoidance compensated for the low heat and desiccation tolerances of these two particular varieties. Jordan et al. (1979) have also demonstrated genetic differences in the pattern of root growth of sorghum grown in hydroponics.

Leaf senescence, which reduces transpiration losses, is well known as a mechanism for drought avoidance. Zartman and Woyewodzic (1979) have shown that root growth and leaf senescence have an association in sorghum. They compared the pattern of root growth in a senescing and a nonsenescing hybrid of sorghum throughout the life cycle under field conditions and found that root density and distribution were similar for about the first 3 weeks after planting. At this stage the nonsenescent hybrid began to establish adventitious roots more rapidly than the senescent hybrid. However by grain-filling, when leaf senescence was greatest, the senescent cultivar had achieved the highest root density. Grain yield was lowest in the senescent variety under the well-watered conditions of the experiment. This suggests that increased drought avoidance by such mechanisms as increased root growth and decreased leaf area can reduce the amount of photosynthetic products available for grain-filling.

Later maturing cultivars are generally more likely to encounter drought stress in the later stages of the life cycle than are early cultivars. Blum et al. (1977) showed that late maturing cultivars had a greater root volume than early maturing cultivars due to a greater number and growth of lateral branches. Heterosis was also expressed by promoted growth and number of lateral roots, which may in part account for the increased drought resistance of many hybrids when compared with line cultivars.

The ability of the sorghum root to adjust osmotically to increasing soil moisture deficit is discussed later in the section on components of water potential.

Leaf Growth, Photosynthesis, and Leaf Senescence

Drought stress can have major effects on the growth of sorghum plants by limiting leaf-area development, by modifying photosynthetic activity and the transport of assimilates, and by influencing the pattern and rate of senescence. The latter in turn provides a mechanism for reducing transpiration. McCree and Davis (1974) showed that 5 cycles of soil water deficit under hot, dry field conditions caused a decline in both the rate of increase of leaf area and final leaf area attained in sorghum. Reduction was due to a decline in the number of epidermal cells, not just a decline in size of cells. Only about half of the 40% reduction in leaf area due to stress could be attributed to actual reduction in area per cell, with the remainder due to reduction in cell number from a decrease in cell division. Leaf expansion stopped at about -17 bar of ψ_w; it was also noted that unlike corn, in which leaves expanded only at night, sorghum leaves expanded at a constant rate during both day and night. They also concluded that, under conditions of considerable drought stress, it is very difficult to separate the effects of water stress from high-temperature stress. In these particular studies leaf temperatures were as high as $41°C$ at the time of maximum drought stress.

Sharp et al. (1979) noted the sensitivity of sorghum leaves to drought stress. They found an appreciable reduction in the rate of leaf expansion, particularly during the day, when leaf turgor was still positive and greater than 1 bar. They assumed that the reduction in leaf turgor in the stressed plants was in part due to an increase in solute potential. Their work indicated that a very small reduction in leaf turgor potential could cause a considerable reduction in leaf expansion. Stout et al. (1978) also noted that very small changes in turgor potential are associated with considerable changes in leaf and stem growth. Distinct and opposite cultivar responses in leaf growth were noted in response to drought stress. In one cultivar (M-35), water stress extended the period of leaf and stem growth, and inflorescence development was delayed. In another cultivar (NK 300), the leaf and stem growth period was shortened, and inflorescence development was advanced by water stress. In both cultivars, total vegetative growth was reduced by water stress in comparison with unstressed plants. Nevertheless, the leaf growth reflected characteristics of tolerance to and avoidance of the drought stress in the respective cultivars.

Studies of the subcellular changes that occur in leaves during increasing water stress (Giles et al. 1976) indicated that amounts of starch in the bundle sheath chloroplasts were much reduced by -14 bar of leaf ψ_w. Swelling of the outer chloroplast membranes and reorganization of the tonoplast to form small vesicles from the large central vacuole occurred by the time a leaf ψ_w of -37 bar was reached. However, unlike maize, there was not complete structural disruption of the tonoplast. It was concluded that maintenance of tonoplast integrity in leaf cells is an important characteristic of drought-stress resistance in sorghum. Slow and rapid effects of water stress on photosynthesis in leaves were compared respectively by monitoring intact and excised leaves (Pasternak and Wilson 1974). Reduced net photosynthesis in the slowly stressed plants was due largely to stomatal closure, in contrast to the leaves exposed to rapid stress where other internal factors were involved in limiting photosynthesis. Rawson et al. (1978) found that diurnal net photosynthesis per unit leaf area in sorghum was 2.3 times greater than soybean under similar conditions of drought. At the same time, transpiration losses were less in sorghum. The sorghum leaf was able to continue photosynthetic assimilation sufficient to compensate the increased respiratory losses during grain-filling that accompanied increased drought stress.

In a study of diurnal changes in photosynthesis and translocation in relation to drought stress, where sorghum was compared with cotton (Sung and Krieg 1979), it was found that photosynthetic rates were reduced in the species as the midday leaf ψ_w reached −14 and −27 bar, respectively. However, sorghum maintained higher levels of photosynthesis and translocation rate when compared to cotton at similar leaf ψ_w values; the rate of change per bar decline in ψ_w was greater in sorghum than in cotton. In both species photosynthetic rates were reduced with increasing water stress prior to any significant change in translocation rates, suggesting that the former is the more sensitive of the two processes (Fig. 63). Even at

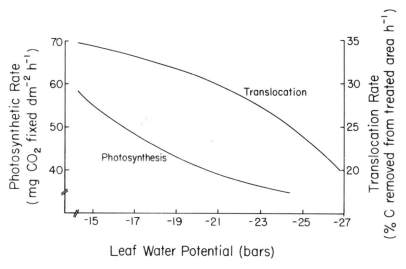

Fig. 63. The relationship between translocation and photosynthesis in sorghum leaves to changes in leaf water potential (after Sung and Krieg 1979).

the severe stress of -27 bar, leaf ψ_w, photosynthesis, and translocation were not completely inhibited in sorghum and this resistance to stress was independent of age. High air temperature, and, in turn, leaf temperature, ultimately limit photosynthesis and interact with water-deficit stress, as indicated in a field experiment reported by Sumayo et al. (1977). Their study of the effects of reduced soil moisture availability on photosynthesis (measured in terms of net carbon dioxide exchange, NCE) indicated that NCE rates were significantly reduced when available soil moisture fell to less than 35% of the maximum. Also, photosynthetic rates declined when leaf temperatures exceeded $33°C$. Modification of leaf temperature by transpiration will be discussed in a later section.

Blum and Sullivan (1972) have shown that there are distinct cultivar differences within sorghum with respect to the way water stress affects photosynthesis, at least with respect to excised leaf sections. Two cultivars known to be fairly drought resistant (M-35 from India and RS 610, a hybrid from the United States) were compared in a desiccation- and heat-tolerance test using leaf discs. Photosynthesis in M-35 was less affected by lowered leaf ψ_w than RS 610. Because of differing abilities of cultivars to adapt and harden from prolonged and increasing levels of drought stress, it was not clear how much of the difference between these two particular cultivars represented the direct sensitivity of the photosynthetic apparatus to drought or a complex of other sensitive characters including stomatal action.

Sorghum, like many other drought-resistant crop plants, shows distinct cultivar differences in the rate and magnitude of leaf senescence, both of which can be influenced by drought stress (Stout and Simpson 1978). Early leaf senescence can be an adaptive response to diminishing soil moisture availability by which leaf area, and thus transpiration, can be reduced (Constable and Hearn 1978). Senescence is manifested as a premature shedding of the lower leaves of the plant (Constable and Hearn 1978; Kannangara et al. 1981; Stout and Simpson 1978; Zartman and Woyewodzic 1979). The process can occur at all stages of the life cycle (Fig. 64) and is particularly marked at grain-filling (Zartman and Woyewodzic 1979) and more pronounced in some cultivars than others (Kannangara et al. 1981; Stout and Simpson 1978; Zartman and Woyewodzic 1979).

Several metabolic products are produced in sorghum leaves as a result of drought stress and accompanying leaf senescence. Large amounts of HCN can be produced as a result of the disrupted metabolism in prematurely senescing drought-stressed leaves. The quantities

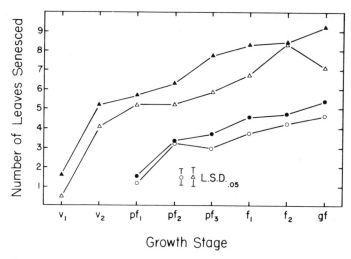

Fig. 64. The number of senesced leaves at different growth stages of two sorghum cultivars, NK 300 and M-35, exposed to drought stress. V = vegetative, pf = preflowering, f = flowering, gf = grain tillering (after Kannangara et al. 1981).

can be great enough to make it dangerous to use sorghum leaves as fodder for cattle (Shukla et al. 1973). Free proline accumulates in drought-stressed leaves of sorghum (Blum and Ebercon 1976; Waldren et al. 1974), and there are distinct cultivar differences in quantities produced. There are small diurnal changes in proline levels primarily related to changes in light intensity (Waldren et al. 1974), and massive accumulations only occur when leaf ψ_w is lowered to around −18 to −20 bar (Blum and Ebercon 1976; Waldren et al. 1974). Free proline accumulation has been significantly correlated with post-stress recovery of leaves, possibly because proline can be a source of respiratory energy for the recovery of the plant (Blum and Ebercon 1976).

Flowering and Grain-Filling

A number of reports indicate the particular sensitivity of the sorghum plant to drought stress through the period when the floral organs are being formed and to a lesser degree when the grains are being filled. Stress during the first of these stages can reduce seed number, during the second stage, seed weight; these together are the major determinants of final grain yield. A comparison of several experi-

TABLE 20: Susceptibility to Drought Stress, Measured as Reduced Final Grain Yield, of the Stages of Growth around Floral Initiation in Sorghum

Stage of Growth	Estimate of % Susceptibility as Loss of Final Grain Yield			
	Musick & Grimes (1961)	Lewis et al. (1974)	Shipley & Regier (1970)	Eck & Musick (1979a)
Late vegetative	32	17	12	—
Boot through bloom	12	34	35	27 to 55
Headings to milk dough	25	—	45	—
Milk dough to soft dough	—	10	6	12

Source: After Lewis et al. (1974).

mental approaches to determine the time of maximum sensitivity to drought stress (Lewis et al. 1974) indicates that the early boot stage, when the florets are being formed and the potential seed number laid down, is particularly sensitive (Table 20). Yield reductions from stress initiated at the early boot stage can come from both reduced seed number and seed size, whereas stress imposed at heading or later only reduces seed size (Eck and Musick 1979a). There are cultivar differences in the manner in which the inflorescence can tolerate or avoid desiccation (Stout et al. 1978; Hultquist 1973). Hultquist (1973) showed that one cultivar could divert photosynthate from the leaves to the lower part of the shoot and roots so that even if the stress was sufficient to cause death of part of the panicle, the remainder, upon rewatering, could survive and continue to develop because some of the plant remained functional. On the other hand, another variety diverted so much photosynthate to the panicle as a result of stress that the remainder of the plant became susceptible to the stress and died. Under the same conditions of drought stress, inflorescence development can be delayed in one cultivar and advanced in another, affecting both floret number and seed weight (Stout et al. 1978). There are important temperature interactions with drought stress that can have highly significant effects on the development of seed number. These are discussed later.

Changes in the Components of Plant Water Status with Drought Stress

Considerable attention has been paid by plant physiologists to the determination of the components of water status in the tissues of sorghum plants exposed at various stages of the life cycle to drought stress. Such parameters as plant ψ_w, ψ_s, and ψ_p can now be measured fairly accurately, even under field conditions, due to the technical advances made in instrumentation over the last few years (Slavik 1974). Interpretation of the ways the determinants of drought stress, such as evaporative demand and diminishing soil water content, induce changes in the energy status and quantity of water in plant tissues has been of major interest to physiologists attempting to characterize the nature of drought resistance in sorghum.

In this section, consideration will be given to recent findings about the way drought stress affects the internal water status of sorghum plants and the manner in which the sorghum plant is believed to adjust to this stress. Stomatal control of transpiration in relation to drought stress will also be considered. Keeping in mind that water potential (ψ), or, more specifically, leaf water potential (ψ_w), is a function of three component potentials—solute (ψ_s), pressure or turgor (ψ_p), and matric potential (ψ_m)—each of which can vary independently in accordance with the equation

$$\psi_w = \psi_s + \psi_p + \psi_m$$

the ψ_w will be discussed first, followed by the components.

Both diurnal and seasonal changes occur in the magnitude of tissue water potential in sorghum. Acevado et al. (1979) monitored leaf ψ_w and relative water content (RWC, a measure of tissue water loss under stress) and found that at any given RWC the leaf ψ_w was lower in leaves at midday than early in the morning due to the increased transpirational losses at midday from increased evaporative demand. This evidence and actual measurements of leaf ψ_s showed that midday ψ_s values were at least 4 bar lower than early in the morning. The decrease in ψ_s was due primarily to an increase in cell sugar concentration, which was considered to represent diurnal osmotic adjustment. This adjustment permitted the maintenance of positive turgor pressure, and thus continued growth, throughout the day, despite increased drought stress. They also concluded that a knowledge of the single parameter ψ_w does not provide sufficient information about the impact of water stress on physiological processes related to water content or cell turgor.

Diurnal changes in leaf ψ_w of sorghum were also recorded by Ackerson et al. (1977) in response to decreasing soil moisture availability and daily changes in evaporative demand. They noted that leaf ψ_w of sorghum came into equilibrium with soil water potentials in the early morning. Adjustment in leaf ψ_s on a diurnal basis permitted increasing turgor pressure, estimated as ψ_p, despite the lowering of leaf ψ_w. They noted a gradient between the top and bottom of the plant in leaf ψ_w which was also noted in another study (Chu and Kerr 1977) in which the upper leaves were found to be considerably more stressed than the lower leaves. The difference between upper and lower leaves could be as much as -7 bar.

Neumann et al. (1974) related leaf RWC to leaf ψ_w and observed that below 100% RWC very small changes in RWC induced considerably larger changes in leaf ψ_w down to a particular critical point, which varied with cultivars, where there was a distinct change. Past this critical point quite large changes in relative water content caused relatively small changes in leaf ψ_w. They concluded that the tensile strength (which is related to the elastic modulus) of the cell walls was involved in the first phase in maintaining cell integrity.

In a comprehensive study (Turner and Begg 1973; Turner 1974) comparing sorghum with maize and tobacco, diurnal and seasonal changes in water potential components were measured along with leaf diffusive resistance as a measure of stomatal activity. Sorghum was able to keep stomata open longer than either maize or tobacco when leaf ψ_w declined. Sorghum was also able to sustain full turgor pressure at a lower ψ_w than either corn or tobacco. Collectively these responses were considered to show a greater adaptation to drought stress by sorghum than by the other two species (Fig. 65).

Seasonal changes occur in tissue water status as the sorghum plant becomes exposed to periods of alternating stress and rewatering, or continued exposure to a pattern of decreasing soil moisture availability combined with increased evapotranspiration potentials. Fereres et al. (1976) found that in a high-radiation environment, leaf ψ_w of sorghum declined during the season to a value of -20 bar. Nevertheless, accompanying decreases in leaf ψ_s enabled the plants to maintain high and positive turgor pressures even at midday, regardless of leaf ψ_w. Leaf extension rate was well correlated with leaf ψ_p. They also showed that if the water stress developed very slowly, the sorghum plant adjusted internal water status so well that ψ_p values were similar to those of plants in well-watered treatments. They noted that interpretation of the water-status parameters can be difficult under field conditions because the relationship

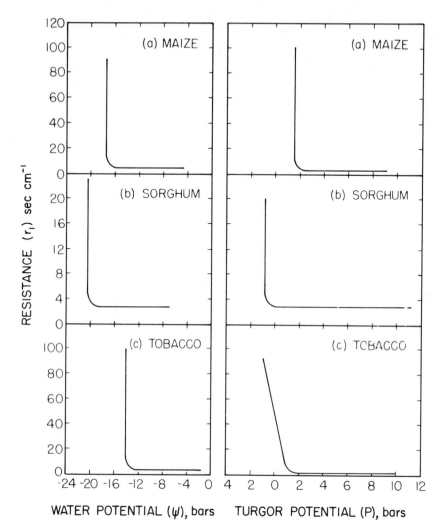

Fig. 65. Relation between leaf resistance (r_1) and leaf water potential (ψ) or turgor potential (P) for maize (a), sorghum (b), and tobacco (c) at various soil water potentials for leaves irradiated at greater than 0.6 cal cm^2 min^{-1} (after Turner 1974).

between RWC and ψ_w changes on both a daily and a seasonal basis. Inuyama (1978a) also noted a seasonal decline in leaf ψ_w in well-watered plants to about -19 bar, whereas drought-stressed plants declined to about -25 to -30 bar. He noted significant varietal differences in leaf ψ_w although the daily and seasonal patterns of change were similar. Stout and Simpson (1978) showed that the seasonal decline in leaf ψ_w was matched by a parallel decline in leaf ψ_s. Significant cultivar differences in leaf ψ_s were apparent by the end of the growing season. When stressed plants were rewatered early in the growing season, the leaf ψ_s returned to values similar to nonstressed plants. However, later in the season the low ψ_s values found in stressed plants did not rise again following rewatering.

This ability to adapt to increasing drought stress in sorghum can be related diurnally and seasonally both to stomatal regulation of water loss and to what has generally been referred to as "osmotic adjustment" (OA). In very drought-resistant crop species, such as sorghum and cotton, OA can be so effective that stomata can remain open even when leaf ψ_w is as low as -28 bar (Ackerson and Krieg 1977; Ackerson et al. 1977). Sorghum is superior to corn in the ability to adjust leaf ψ_s under similar regimes of environmental stress (Ackerson and Krieg 1977). Genotypic differences have been observed among cultivars in the ability to show OA at each growth stage of the life cycle (Ackerson et al. 1980). Stout and Simpson (1978) also noted cultivar differences in leaf ψ_s among nonstressed as well as drought-stressed plants grown under field conditions, which was ascribed to a drought-avoidance mechanism achieved by the accumulation of osmotically active cellular solutes.

Jones and Turner (1978) examined in considerable detail the mechanism of OA reported earlier from the same laboratory for a single cultivar of sorghum. They compared two genotypes of sorghum, RS 610 and Shallu, generally considered to be fairly drought resistant. They found no differences between the cultivars in degree of osmotic adjustment or in other components of leaf tissue water relations such as ψ_w, ψ_s, or ψ_p. They concluded that the superior responses of RS 610 over Shallu in drought-tolerance tests applied by other workers must be due to differences in other factors such as rooting habit, conductance of water in the xylem, or desiccation tolerance. Nevertheless, they demonstrated diurnal and seasonal OA in both cultivars. The stress treatments altered the relationship between leaf ψ_w and RWC so that previously stressed plants were able to maintain higher tissue water contents than control plants at the same ψ_w. Althouth OA permitted prestressed plants to maintain a higher ψ_p at a given leaf ψ_w than control plants, zero ψ_p still occurred at approximately the same value of RWC (94%) irrespective of previous stress history. They noted that stress preconditioning lowered tissue elasticity which increased the value of the volumetric elastic modulus. The relationship between osmotic adjustment and the water potential at which sorghum stomata close has been studied by Turner et al. (1978).

A later paper (Jones et al. 1980) demonstrated that the increase in osmolarity which decreased ψ_s when diurnal OA occurred was almost entirely due to increases in sugars and total inorganic ions, in approximately equal proportions. Readjustment of leaf ψ_s following a decrease in drought stress—for example, by rewatering—can occur slowly (6 to 11 days) (Jones and Rawson 1979) whereas recovery of both net photosynthesis and leaf conductance can be quite rapid (less than 3 days).

There is a scarcity of information on root tissue water status, primarily because of the practical difficulties of removing roots from soil without damage. Roots can be grown in a hydroponic medium, stressed with osmotica, and can be removed for water status measurements, but it is not clear how these conditions relate to a normal soil. In soil with a diminishing water content, matric potential of the soil (ψ_m) may be a very significant component compared to soil ψ_s, and the total ψ of the soil will determine the resistance to water movement into the plant. Cruz-Romero and Ramos (1979) measured hydraulic root resistance (R_t) per unit root length and per unit leaf area as transpiration rate changed in relation to a declining soil water availability: $R_t = (\psi_s - \psi_x)/F$, where ψ_s is the soil water potential and ψ_x the xylem sap water potential in the lower part of the plant; F is the rate of water loss from the entire plant. They found a significant linear decrease in root resistance with increasing transpiration rate. In stressed plants the hydraulic root resistance per unit leaf area was nearly twice as high as in nonstressed plants at low levels of transpiration, indicating considerable control over the passage of water through roots after drought stress.

A number of investigations have been focused on the mechanisms that can increase leaf resistance to water loss, which in turn lowers transpiration losses as atmospheric evaporative demand increases or as soil moisture reserves are depleted. Stomatal action and the role of such factors as hairiness of leaves, leaf orientation, leaf rolling, epidermal waxiness, and osmoregulation have all been investigated in relation to specific changes in leaf diffusive resistance and conductance, transpiration, and photosynthesis. These factors will be considered in the remainder of this section.

Blum and Sullivan (1974) have shown that when sorghum plants are given consecutive short periods of drought stress followed by rewatering cycles, stomata are less sensitive in successive cycles to closing under the influence of stress. Leaf ψ_w was lower at a given soil ψ_s when stomata closed following prestress. They noted that in plants not previously stressed the stomata were sensitive to any initial reductions in leaf ψ_w, but once closure was initiated (at about -8 bar) the stomata remained only partly closed (about one-quarter to one-third of the aperture) even when the leaf ψ_w dropped to -19 bar, indicating that during this second phase transpirational control was primarily nonstomatal. Fereres et al. (1978) also indicated that under field conditions where plants were progressively stressed during the season, there was no stomatal closure due to water stress even when leaf ψ_w reached -20 bar in older plants. In younger plants under the same conditions where water stress was induced rapidly, stomata shut at -14 bar. In spite of the decrease in ψ_w combined with open stomata throughout the season, turgor was maintained in the tissues by means of a corresponding drop in leaf ψ_s. Thus the maintenance of cell turgor was seen to be strongly associated with open stomata.

Inuyama (1978a) indicated cultivar differences within sorghum in the threshold level of ψ_w at which leaf diffusive resistance (R_L) increased sharply due to stomatal closure. The rate of development of stress can also influence the degree of stomatal aperture control (Jones and Rawson 1979), thus affecting photosynthesis as well as transpiration. In their experiments, rapid increases in drought stress induced the lowest rates of net photosynthesis and leaf conductance with essentially no adjustment in leaf ψ_s, indicating that stomatal closure was the principal mechanism for increasing leaf resistance to transpirational water loss. On the other hand, slower rates of buildup of stress permitted osmotic adjustment and higher rates of net photosynthesis and leaf conductance. They noted that if stress progressed slowly, a prehistory of stress had no significant influence on the response to a new stress cycle. A significant conclusion from this study was that the concept of a switch-like action of stomata at some threshold leaf ψ_w was not valid under conditions of slowly imposed drought stress—conditions commonly found under field conditions. They attributed the differences commonly found between laboratory-grown plants in small containers, which undergo rapid stress as soil dries out, and field-grown plants, which develop stress slowly, to the different patterns of stomatal action and solute accumulation. They also concluded that osmotic adjustment alone cannot be the total explanation for the high values of leaf conductance at low leaf ψ_w in slowly stressed plants since there were similar levels of solute accumulation and similar relationships between leaf ψ_s and leaf ψ_w in plants stressed more rapidly but which showed different relationships between leaf conductance and leaf ψ_w. They suggested that, possibly, guard-cell solute accumulation, rather than total leaf tissue solute accumulation, could be the integrating mechanism between cell water status and transpirational control.

McCree (1974) suggested that the ability of stressed sorghum plants to show consistently higher leaf conductance (and thus photosynthesis) at a given leaf ψ_w than nonstressed plants is a drought-avoidance mechanism. The stressed plant actually transpires more water and accumulates more carbon and thus has greater root growth to secure water and survival than a nonstressed plant. Pasternak and Wilson (1976) showed that this avoidance characteristic can occur to differing degrees in different parts of the plant. They noted that when severe water deficits reduced the photosynthetic rate in leaves, the photosynthetic efficiency in the heads increased from about 12% of the total, in a well-watered plant, to about 88% under severe stress. This was accompanied by a proportional increase in transpiration from the heads. In addition Turner and Begg (1973) showed that the upper leaves on the sorghum plant had lower stomatal resistances than the lower leaves due to a higher ψ_s in the former. The differences were maintained at all values of ψ_w.

In a recent paper, Ackerson et al. (1980) indicated that there is genetic variation within sorghum in the degree of transpirational control during the life cycle. This variation depends on changes in the proportion of "water-spending" and "drought-tolerating" aspects of their physiological response to drought stress. They observed that prior to flowering there were discernible genotypic differences in the leaf water potential required to initiate stomatal control of transpiration and also in the rate of change of conductance per unit change in leaf ψ_w. However, there were no significant differences among genotypes in conductance when leaf ψ_w was high. After flowering, conductance remained high, even when leaf ψ_s declined to -26 bar, due to the ability to maintain turgor by osmotic adjustment. In this way, they suggested, water conservation is practiced before reproductive development starts, but after flowering photosynthetic productivity is maintained at the expense of water conservation.

Stomatal control in rapidly stressed plants may be related to the endogenous production of the plant hormone cis-abscisic acid and to trans-farnesol. Both of these substances cause stomata to shut rapidly if applied exogenously, and they are produced endogenously in large quantities in stressed leaves (Mansfield 1976a).

The considerable ability of the sorghum plant, compared to such crops as maize and soybean, to maintain a positive turgor pressure through the season, even under stress, by a combination of osmotic adjustment and control of stomatal aperture ensures a high water-use efficiency (WUE). Teare, Kanemasu, et al. (1973) found that the WUE (gm DM/kg H_2O) of sorghum was approximately 3 times that of soybeans whether compared on a total dry-matter or grain-yield basis. WUE appeared to be unaffected by the slow application of drought stress (Jones and Rawson 1979); however, as the rate of application of stress increased, WUE declined. WUE was more sensitive to changes in leaf ψ_w at grain-filling than at other periods of growth. From a comparison of 5 other crop species with sorghum, Rawson et al. (1977) concluded that it is difficult to assign a comparative short-term value for WUE to each species because of the way WUE changes with age of leaves. In this experiment, carried out under conditions of adequate soil moisture, transpiration in each crop increased linearly with increasing vapor-pressure deficit. Sorghum transpired considerably less water than either soybean, sunflower, or wheat at any given level of vapor-pressure deficit.

Transpiration is evidently important in temperature control of the leaf (see next section). Sumayo et al. (1977) confirmed an earlier report (Kanemasu et al. 1976) that in well-watered sorghum plants there was a critical air temperature of $33°C$ below which the leaf temperature was higher than air temperature. Above $33°C$ the transpiration rate increased so that leaf temperature remained below ambient. By contrast, in drought-stressed sorghum there was a reverse situation. Below a crossover temperature of $35°C$ the leaf was cooler than the air, while above this temperature the leaf was warmer than the air.

A strategy for reducing water loss from the plant is to reduce radiation intercepted by the canopy. Sorghum plants generally have considerable quantities of epicuticular wax on stems and leaves. Blum (1979) has demonstrated that the wax causes reduction in net radiation of the canopy as well as a reduction in cuticular transpiration and an improvement of stomatal control over transpiration. WUE can also be improved by wax deposition (Chatterton et al. 1975). Active and passive leaf movements and pubescence may also reduce net radiation interception, but there is little evidence about these avoidance mechanisms in sorghum.

In summary, then, it can be seen that the sorghum plant has several mechanisms for adjusting to water deficits in the environment, whether they occur slowly or rapidly. The net effect of these adjustments in the components of water potential is to maintain cell turgor pressure and permit continued photosynthesis and growth despite a considerable degree of drought stress. In an evolutionary sense, the high degree of osmotic adjustment represents an improved tolerance to high osmotic potentials in a manner similar to that achieved by some of the true xerophytes.

Temperature Interactions with Drought Stress

Both high and low temperatures can influence the net reaction of sorghum plants to drought stress. High radiation input will induce a high evapotranspiration potential as well as a high ambient air temperature. Sorghum has a higher optimum temperature requirement for kernel development than either maize or wheat (Chowdury and Wardlaw 1978), indicating its adaptation to high temperatures. There are genotypic differences among cultivars of sorghum with respect to the degree of heat tolerance (Troughton et al. 1974). Drought stress reduces the optimum day and night temperatures for both vegetative growth and grain-filling in sorghum (Tateno and Ojima 1976). Under moderate soil water conditions, the most favorable temperature for grain production was over the day/night range of 35 to $25°C$ and 25 to $15°C$. Under drought conditions, the lower value of 25 to $15°C$ was nearer the optimum.

Both low night temperatures ($5°$) and high night temperatures ($30°$) compared to an intermediate temperature, reduced the extent and rate of stomatal opening the following day in sorghum (Pasternak and Wilson 1972). Photosynthesis was correspondingly reduced. The reduction in rate of stomatal opening was attributed to the development of water deficits in the leaves. High night temperatures at panicle initiation severely reduced subsequent grain yields due to a reduction in potential grain number (Eastin et al. 1975). A night temperature as little as $5°$ above optimum reduced subsequent grain yield as much as 36%. This temperature effect draws attention to the great sensitivity of panicles, during early development, to temperature changes as well as the responses to water stress described in the previous section. The temperature reductions in seed number may well have their origins in subtle and rapid changes in the water status of panicle tissues which reinforce the direct effects of temperature on metabolic events.

Sullivan and Ross (1979) have reviewed the association between heat-stress and desiccation- (drought-) stress tolerance in sorghum and have indicated that in general there is a positive correlation between the two parameters. They found a significant positive correlation between heat tolerance and yield. Because a number of workers have found similar correlations in other crops, there has been a general interest in using heat stress to select for drought resistance (Sullivan and Ross 1979). Sullivan (1972) devised a method for screening for heat tolerance in sorghum that can be used for selection in plant breeding programs. Considerable genotypic variation for heat and drought tolerance has been demonstrated in sorghum with this method. Some of the effects of heat stress are probably due to the negative effects of high temperature on photosynthesis, which can be separated from the indirect effects of stress on stomatal closure (Sullivan and Ross 1979).

Growth Regulators

Naturally occurring plant growth regulators are believed to play an important role in all aspects of plant growth and development. Until recently, the technical difficulties associated with their identification and quantitative analysis limited investigation of their role in the control of growth changes associated with drought stress in any plants. However, recent advancements in multihormone analysis (Simpson et al. 1979) have now made it possible to examine hormone changes in sorghum in relation to the response of plants to drought stress. Analysis of auxins, abscisins, and cytokinins (Dunlap and Morgan 1978; Durley et al. 1978; Kannangara et al. 1978) by high-performance liquid chromatography has demonstrated the presence of these families of growth regulators in sorghum. Isolation of gibberellins in sorghum has also been reported (Dunlap and Morgan 1978).

Both farnesol (Fenton et al. 1977) and abscisic acid (Mansfield et al. 1978) can control stomatal opening in sorghum when applied to leaves, an effect that can persist for several days in the case of farnesol. Larque-Saavedra and Wain (1976) showed an association between the levels of endogenously produced abscisic acid in wilted leaves and a range of drought-tolerant genotypes in sorghum. Durley et al. (1981) have shown that both diurnal and seasonal fluctuations in the levels of abscisic acid in leaves, while indicating genotypic differences between cultivars, are primarily related to changes in leaf ψ_w induced by the environment. As leaf ψ_w decreases with an increase in environmental stress, abscisic acid levels increase substantially.

Gibberellin applied to sorghum hastens floral initiation (Williams and Morgan 1979) and inhibits root growth (Bhatt et al. 1976), aspects of growth which are similarly affected by drought stress. On the other hand, gibberellin increases the height of many sorghum cultivars and thus acts in the opposite manner to drought stress, which reduces height (Morgan et al. 1977).

The evidence gained to date suggests that hormones may be significant intermediaries between the growth response of sorghum and environmental changes, such as drought stress, that modify growth.

Modeling Sorghum Growth

There has been an increasing interest in recent years in developing models of sorghum growth, for several reasons. Sophisticated satellite (Nixon et al. 1976) or aircraft scanning systems for monitoring canopies of crop plants were developed in the 1970s. It is now possible to retrieve both light (Blum 1979; Collins 1978) and radar (Bush and Ulaby 1978) images of crop canopies for particular ranges of the electromagnetic spectrum. Correlating image information with specific aspects of plant growth requires a good dynamic model of the growth of the sorghum plant throughout its life cycle under a range of environmental conditions. Such a model can only be developed on the basis of base-line data obtained from field or controlled-environment conditions. Basic information such as leaf area (Lomte et al. 1979; Van Arkel 1978) and tissue water status are of particular significance in relation to the absorption and reflection of specific wavelengths of radiation. Knowledge of the changes in plant water status due to changes in the environment, such as drought stress, are essential to the interpretation of imagery—for example, in relation to the absorption and reflection of infrared radiation (Blum 1979; Heilman et al. 1976).

Another area in which modeling has significance is the use of meteorological data (Blad et al. 1978) to predict final crop yields based on a periodic analysis of changing environmental parameters such as rainfall, irrigation, temperature, and evapotranspiration fluxes (Verma and Rosenberg 1977). Interpretation of the effects of these changes depends on an understanding of the interrelationships between the soil-plant-aerial environment systems. Modeling plant growth and its relationship to the environment is of considerable value to crop physiologists because it demonstrates where existing gaps in physiological knowledge occur. This encourages research to complete a more holistic and useful understanding of the functions of the sorghum plant.

One of the earliest models was developed to predict grain yield under different climatic conditions in relation to irrigation of

sorghum (Hanks 1974). The development of models has evolved
from accumulating information about the subparts—for example,
roots (Hewitt and Dexter 1979; Van Bavel and Ahmed 1976) and the
leafy canopy (Arkin et al. 1978; Ritchie and Arkin 1976; Vanderlip
and Arkin 1977)—to comprehensive models that attempt to consider
all aspects of the growth cycle in relation to dynamic changes in the
soil water status and aerial environment (Arkin et al. 1976; Hodges
et al. 1979; Slabbers et al. 1979). The model of Hodges et al. (1979),
for example, incorporates the reductions in photosynthesis due to
the effects of high temperatures and drought stress.

The acquisition of reliable data for developing effective models
requires the measurement of a considerable number of parameters at
the same point in time, repeated at many intervals of time under a
range of environmental changes. This can be difficult and expensive
in terms of instrumentation and manpower, and it also takes a
number of seasons of field data as well as data from controlled-
environment situations. Nevertheless, the potential returns in terms
of agronomic practice and physiological understanding are high.
There is clearly a need for more of this synthetic type of research
to produce the basic models of sorghum growth upon which the
effects of drought stress can be superimposed. The full dimensions
of the effects of drought stress on sorghum plants will then become
apparent.

Breeding and Selection for Drought Resistance

As physiological knowledge about the nature of drought resistance in
sorghum has increased, interest in breeding and selection for drought-
resistance characteristics has also increased. Such components of
drought resistance as avoidance (root growth, osmotic adjustment,
stomatal activity) and tolerance (sustained photosynthetic activity,
desiccation and heat tolerance) all show some genotypic variation, and
therefore they have potential use in breeding and selection programs.
Blum (1979) has clarified the nature of the problem for sorghum by
showing that there are two distinct avenues or philosophies for
approaching the problem of breeding for drought resistance. The
first, governed by the need to improve crop yields, accepts that a
superior yielding variety under optimal conditions will probably
also yield relatively well under suboptimal conditions. The second
maintains that higher potential yield is irrelevant and that what is
important is to breed for adaptation to a specific environment such as
the semiarid zone. That is, selection for a high genotype–environment

interaction. Yield is a complex character determined collectively by a wide range of factors throughout the life cycle of the plant. Nevertheless, yield potential and yield stability can be separated and are characteristics largely independent of each other. Drought resistance would be an important component of the stability of yield in a drought-prone environment. For these reasons, Blum suggests that breeding for yield and drought resistance can be handled separately along much the same lines as past experience in breeding for disease resistance. Each individual drought-resistance component will have to be evaluated for its degree of independence from association with yield. He has suggested that even a negative association between potential yield and drought-resistance components need not exclude the resistance component since, even under the most favorable environment, plants will always be exposed to some form or degree of environmental stress. What is important is to obtain some, rather than no, yield under extreme drought stress and also to obtain a range of increasing yield potentials suited to a range of environments between extreme aridity and plentiful irrigation.

Turner (1979) points out that there are basically 3 types of drought-resistance plants. Thus breeding would have to deal with one or a combination of these qualities, that is, drought escapers, drought avoiders maintaining high tissue water potentials, and drought tolerators maintaining low water potentials. Thus the possible objectives for a breeding program to incorporate drought resistance combined with a reasonable grain yield in sorghum have now become clearer. However, the means of incorporating desirable drought-resistance characteristics in a plant is still far from clear.

This inability to construct a more effective plant for a specific situation is related to the lack of sufficient suitable genetic markers highly correlated with each attribute—for example, tolerance or avoidance. Nevertheless, the number of useful markers has been steadily increasing (Martin 1977). Collectively, these markers will establish the sieve that can screen sorghum populations, initially for some of the components of drought resistance and eventually for most of them. As with the slow but steady progress in increasing yields of crop plants, breeding for drought resistance will not be achieved by a single-factor approach.

Some examples of recent advances in selection of specific traits by the application of new screening methods are outlined in the remainder of this section. Blum (Blum 1975a; Blum et al. 1978) has developed an approach to selection of dehydration avoidance in sorghum plants growing under field conditions by monitoring infrared photographic images of plots using aerial photography. Plants that are dehydrated absorb less infrared radiation. In another approach the same author has demonstrated genotypic differences in proline levels in leaves of stressed plants. These proline levels are positively correlated to poststress recovery and can be used for genetic selection. Genetic variability has been confirmed both for stomatal density (Liang et al. 1975; Suh et al. 1976) and stomatal sensitivity to water stress (Henzell et al. 1975, 1976). Techniques for monitoring these characteristics have been established. Selection against non-stomatal transpiration losses, which can be reduced by the presence of epicuticular wax (Blum 1975b; Sanchez-Diaz et al. 1972) can be carried out by a rapid colorimetric method using an acidic $K_2Cr_2O_7$ reagent (Ebercon et al. 1977).

Simple heat- and desiccation-tolerance tests using small pieces of sorghum leaf have been developed by Sullivan (Sullivan 1972; Sullivan and Ross 1979). The tests have indicated significant genotypic variation in both of these characteristics in sorghum. Cultivar differences have been established in tissue water retention, protein level, and chloroplast activity in relation to drought resistance (Kushnirenko et al. 1973) and also to leaf water potential and leaf diffusive resistance (Inuyama 1978a). Each of these attributes can now be monitored by simple techniques. Thus the future prospect of finding new techniques for identifying specific traits related to the components of drought resistance in sorghum seems encouraging.

Field Approaches to Improve Drought-Stress Resistance

Blum (1972) and Blum and Naveh (1976) have demonstrated that, under conditions of limited soil water availability with rainfed conditions, significant improvement in yield and water-use efficiency occurs if the density of planting is increased. The increased plant competition hastens the life cycle so that the extreme drought stress at the end of the growing season is avoided. Accurate timing of the planting date under these conditions is significant. McCauley et al. (1978) have indicated that field geometry, in this case orienting rows in a north–south direction in a specific row spacing (45 cm apart), can diminish evapotranspiration compared to other arrangements, thus helping to avoid drought stress. Stanhill (Moreshet et al. 1977; Stanhill et al. 1976) has investigated the use of kaolin sprays to increase foliage reflectance of incoming radiation. Although the kaolin increased leaf senescence in the whitened leaves, there was only a minor reduction in leaf diffusion resistance such that a 26% reduction of solar radiation reduced net photosynthesis by 23%. Nevertheless, grain yields were increased, indicating a substantial gain from better water status and the reduction of respiratory losses by lowered leaf temperatures. Attempts have also been made to modify the response of sorghum plants to high temperatures by the use of compounds such as substituted pyridazinone which affect membrane integrity and permeability (St. John and Christiansen 1977). Modification of soil temperature by the use of straw mulches (Unger 1978) reduced germination rate due to soil cooling and slowed development of plants; however, later in the season, when soil moisture became limiting, the mulched plants grew better than plants on bare soil due to the better moisture conditions and reduced drought stress.

Water-use efficiency is important under irrigation as well as under conditions of drought stress. Szeicz et al. (1973) has calculated that, under irrigation conditions in Texas, doubling crop density— under existing conditions of maintaining stomatal resistance near minimum—could increase yields by about 100%, whereas evapo-transpiration would only increase by about 33%. This could represent a significant improvement in water-use efficiency. Most of the voluminous research to find optimal planting density on different soils and in different geographical zones of the world indicates a strong relationship between planting density and geometry and crop water-use efficiency.

The paucity of practical field techniques for reducing the effects of drought stress in sorghum, other than the use of irrigation, high-lights the relative importance of the breeding approach to building drought resistance into the plant rather than a manipulation of the environment to achieve optimum plant growth and yield. Sorghum clearly has considerable genetic variation in those aspects of the plant that confer adaptation to drought stress. These attributes can be recombined in new, fruitful combinations—provided that plant physiologists and plant breeders can work together and have a common understanding of the agronomic and environmental situation they are designing the plants for. There is every reason to conclude that substantial advances can be made in the near future in enhancing the drought-resistant properties of sorghum.

VICIA FABA

Vicia faba L. is a valuable crop plant. It can be used as an effective break crop in rotations of continuous cereal cropping, while, at the same time, it produces seeds with high protein content (between 20 to 25%). Despite these facts, the crop is not widespread, because of its unreliability in seed production. The average seed yield in Britain is around 3,000 kg ha^{-1} (Smith and Aldrich 1967). However, yields in experimental plots were as high as 6,300 kg ha^{-1} (Ishag 1969) with possibilities of reaching 9,000 kg ha^{-1} (Sprent et al. 1978), which is indicative of the potential of the crop. Pod-shedding is regarded as the main source of the fluctuations in yield. The fluctuations may arise from either physiological (Sprent et al. 1977) and/or environmental factors (Ishag 1969). From 1968 onward an effort was made in Britain (mainly at the Universities of Reading and Nottingham) to study the effects of major environmental factors on the growth and productivity of the *Vicia faba* L. crop. Thus the effects of temperature and solar radiation were studied in detail (Bull 1968; Dennett et al. 1978). At the same time, the effect of soil water stress on internal water balance and crop growth was extensively studied at the University of Reading (Kassam 1971; Karamanos 1976).

In this section, results from the research progress on drought-stress physiology of *V. faba* L. for both the broad bean and the field bean sections of the species will be presented. Most of the research refers to field beans (*V. faba minor* cv. Maris Bead) and comes from the University of Reading (Department of Agricultural Botany), but results from other sources will also be provided. In the first part of this section the factors influencing plant water relations will be examined (tissue characteristics, transpiration, water absorption) while in the second part some effects of drought stress on plant growth and physiology will be presented.

Water Relations

Tissue Water Relations

Approach   To study the tissue water relations of field beans, the osmometer approach was adopted. According to that approach, the effect of the matric forces on tissue water status is negligible for the range of water contents encountered in living plants; accordingly, ψ_m was taken as zero. Thus the tissue water potential (Ψ) is regarded as the sum of the solute (ψ_s) and pressure (ψ_p) potentials. In general, the osmometer approach is regarded as valid only when considering fully grown tissues of mesophytes where cell matrix (colloids and cell-wall material) is restricted. On that basis, fully grown field bean leaves which consist of thin-walled and fully vacuolated mesophyll cells can be regarded as behaving in an osmometer-like fashion. The lack of departure from an ideal osmotic behavior of the bean leaves (Fig. 66) verifies the theoretical considerations about the absence of matric effects.

A better understanding of the physiochemical factors controlling tissue water relations was achieved by using models containing some parameters associated with tissue characteristics (Philip 1958; Gardner and Ehlig 1965; Warren Wilson 1967a, b, c). In field beans, a model

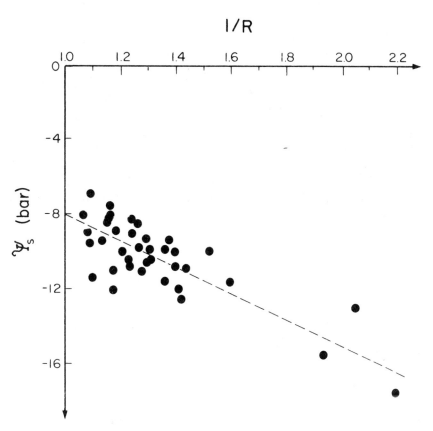

Fig. 66. The relation between the reciprocal of the relative water content $(1/R)$ and the solute potential of the youngest fully expanded leaf (ψ_s) in broad beans. ψ_s was determined with a thermocouple psychrometer on frozen and thawed leaf tissue. The broken line indicates the fitted linear regression ($r = -0.8$). The linearity suggests that the law of van't Hoff (Eq. 47) holds satisfactorily, and matric effects are probably absent for the given range of R (after Karamanos 1978b).

with 2 variables and 3 parameters was used for the study of water relations (Kassam and Elston 1974; Elston et al. 1976; Karamanos 1978b):

$$\Psi = \frac{\psi_{so} R_o}{R} + \epsilon'(R - R_o)$$ (45)

where ψ_{so} is the solute potential at zero turgor, R_o the relative water content at zero turgor, and ϵ' the slope of the linear regression between the pressure potential and the tissue relative water content (R). The first term of the right-hand side of Eq. 45 expresses ψ_s and the second one ψ_p. The 3 parameters ψ_{so}, R_o, and ϵ' are related to chemical and physical properties of the tissues. Thus, ψ_{so} depends on the solute content of the tissue when $\psi_p = 0$; the higher the solute content at this reference stage, the lower the ψ_{so}. The parameter ϵ' reflects the elastic properties of the cell walls because it is proportional to the elastic modulus or coefficient of enlargement of the walls: the higher the ϵ', the faster the drop in cell turgor for a given amount of water lost, namely, the lower the elasticity of the cell walls. Finally, the parameter R_o, which represents the volume of the cell available for water when $\psi_p = 0$, depends on both the solute content at zero turgor (ψ_{so}) and the elasticity of the cell walls (ϵ') (Karamanos 1978b):

$$R_o = \frac{\epsilon'}{\epsilon' + \psi_{so}}$$ (46)

The higher the R_o, the smaller the amount of water that cells can lose before their turgor drops to zero.

The derivation of the model was based on two assumptions. First, that the law of van't Hoff is valid. This is a consequence of the ideal osmotic behavior of the cells. Accordingly:

$$\psi_s R = \psi_{so} R_o = \text{constant}$$ (47)

and, hence,

$$\psi_s = \frac{\psi_{so} R_o}{R}$$ (48)

The assumption was experimentally tested by plotting ψ_s against $1/R$ in broad bean leaves, and it was found to hold satisfactorily (Fig. 66). The second assumption was that ψ_p is linearly related to

TABLE 21: The Linear Regressions of the Pressure Potential (ψ_p) against Relative Water Content (R) for Field Bean Plants Growing under Wet (W), Dry (D), and Intermediate (M) Soil Water Conditions

Treat- ments	1974		1975	
	Regression	r	Regression	r
W	$\psi_p = -1.6 + 2.4R$	0.85	$\psi_p = -0.8 + 1.3R$	0.76
M	$-1.0 + 1.7R$	0.75	$-0.6 + 1.0R$	0.67
D	$-0.8 + 1.4R$	0.90	$-0.8 + 1.3R$	0.87

Source: After Karamanos (1978b).

the change in cell water volume down to zero turgor ($R - R_o$):

$$\psi_p = \epsilon'(R - R_o) \tag{49}$$

where ϵ' is the slope of the linear regression. This implies that the cell wall has elastic properties and obeys Hooke's law. The validity of this second assumption is questionable because several investigators found that the relation between ψ_p and R was curvilinear (Haines 1950; Gardner and Ehlig 1965). Nevertheless, experimental evidence with field beans (Elston et al. 1976; Karamanos 1978b) suggests that the assumption of linearity holds satisfactorily for plants that are not experiencing drastic changes in their water status (Table 21).

Some of the variables and parameters of the proposed model can be measured directly, while others are derived from calculations. Thus the relative water content (R) can be easily measured, according to Barrs and Weatherley (1962). The use of the length-change technique for the measurement of Ψ (Kassam 1972) offers considerable advantages for the evaluation and calculation of the various parameters. According to this method, the change in length of leaf strips allowed to equilibrate with mannitol solutions of different concentrations is plotted against the osmotic potential of the corresponding solutions (Fig. 67). Apart from the measurement of Ψ, this technique allows a direct measurement of ψ_{so} and a calculation of R_o from the changes in the length of the strips. The solute potential is calculated from Eq. 48 while the pressure potential is found by subtracting ψ_s from Ψ. Finally, ϵ' is calculated from the equation:

$$\epsilon' = \frac{\psi_p}{R - R_o} \tag{50}$$

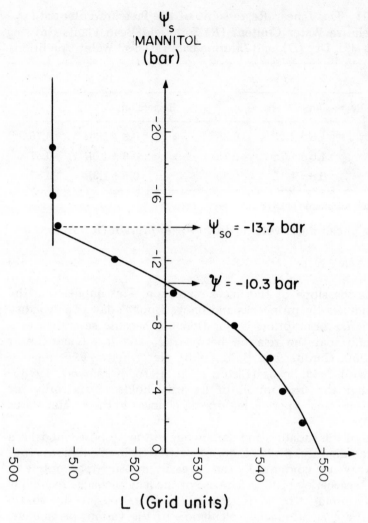

Fig. 67. The determination of leaf water potential (ψ) and solute potential at zero turgor (ψ_{so}) with the length-change technique (Kassam 1972). The length of leaf strips (expressed in arbitrary grid units) after equilibrium with mannitol solutions is plotted against the ψ_s of the corresponding solutions (data from Karamanos 1976).

Effects of Leaf Age and Position   The changes in the variables and parameters of leaf water status with leaf age and insertion level were extensively studied by Kassam (1971) and Kassam and Elston (1976), using the results from a field experiment in the spring of 1968. Leaf samples were taken at 3- to 4-day intervals throughout the season. In general, an acropetal decrease in both Ψ and R was found on most sampling occasions (Fig. 68).

These patterns are not surprising, because of the increased evaporative load that the upper leaves receive. Thus the lowering of Ψ from the bottom to the top of the plant creates a gradient that

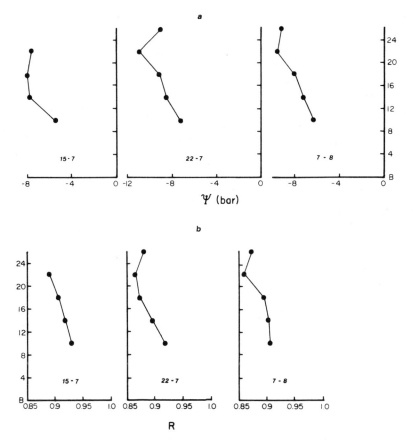

Fig. 68.   The values of variables of plant water status (profiles) at different leaf positions on field bean plants on three different sampling occasions (15/7, 22/7, and 7/8/1968). Leaves are numbered from the bottom (B) of the plant. (a) Water potential (ψ). (b) Relative water content (R) (adapted from Kassam and Elston 1976).

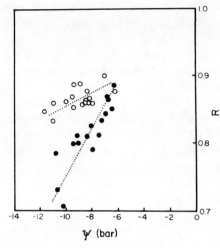

Fig. 69. The relation between the water potential (Ψ) and relative water content (R) for leaves at high (younger, filled circles) and low (older, open circles) positions on field bean plants (after Kassam and Elston 1976).

sustains water flow through the plant. However, the magnitude of the change in Ψ for a given change in R differed with leaf position: the lower the leaf position, the steeper the slope of the relationship between Ψ and R. (Fig. 69). This results in a more abrupt fall in Ψ for a given amount of water lost in the lower leaves. Thus the upper leaves, which receive a much higher evaporative load, are able to withstand much greater dehydration than the lower ones before their water potential falls to the critical level of turgor loss.

The profiles of ψ_{so}, R_o, and ϵ' in the plant may help to understand the origin of the different relationships between Ψ and R in the different leaf positions. In general, ψ_{so} showed a consistent acropetal decrease on the plant (Fig. 70a). The differences between ψ_{so} among leaf positions were more pronounced as the crop aged. An acropetal decrease was also found when examining the profiles of both R_o and ϵ' (Figs. 70b and c).

The higher values of ψ_{so} in the lower leaves do not explain the steeper relationships between Ψ and R in these leaves. Alternatively, the higher values of both ϵ' and R_o imply less-elastic cell walls and restricted possibilities for water loss in the lower leaves. Consequently, the tendency of the lower leaves to wilt first (Kassam 1975) can be attributed to two sets of factors: (1) to the higher values of both ϵ' and R_o which cause the more rapid fall of the leaf turgor to zero, and (2) to the higher (less negative) ψ_{so} which itself implies a quicker wilting. The term wilting refers to physiological or "true" wilting which occurs when $\psi_p = 0$, in contrast with the "apparent wilting" which refers to the situation when leaves droop under their own weight but their turgor is still above zero (Kassam 1975) (Fig. 71).

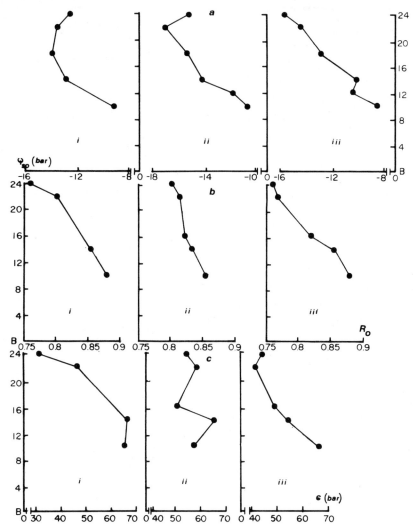

Fig. 70. The profiles of some parameters of plant water status on field beans on three different sampling occasions (i: 15/7, ii: 22/7, iii: 7/8/1968). (a) Solute potential at zero turgor (ψ_{SO}). (b) Relative water content at zero turgor (R_O). (c) Slope of the linear regression between the pressure potential and relative water content (ϵ') (adapted from Kassam and Elston 1976).

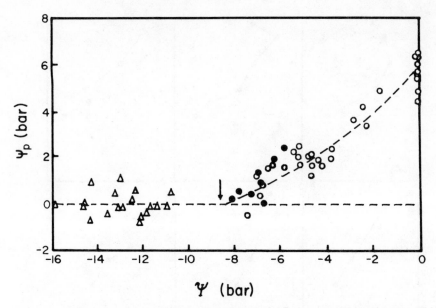

Fig. 71. The relation between the water potential (Ψ) and the pressure potential (ψ_p) of glasshouse-grown broad bean plants for turgid ($\circ$), drooped ($\bullet$), and buckled ($\triangle$) leaves. The arrow shows the point when ψ_p first becomes zero (after Kassam 1975).

Kassam (1975) found that in addition to the loss of water through transpiration, considerable water is lost from the lower leaves via a direct transport to the higher (younger) leaves during periods of water stress. This sink activity of the lower leaves is more apparent when it is considered that they contain more water than the upper leaves both as a whole and per unit leaf area. The changes in both R and Ψ with age did not show any consistent trend and appeared to be influenced by the soil and aerial environment (Fig. 72).

However, the patterns of the parameters ψ_{so}, R_o, and ϵ' were influenced to a different extent by leaf age. Thus the direction and magnitude of change with age in ψ_{so} (Fig. 73a) of a leaf depended on the time it emerged in the life of the plant. For example, earlier leaves (leaf 12B) showed a different pattern in comparison with leaves that emerged later in the season (leaf 22B). Despite a sudden fall at the end of July, which was common for all leaves, R_o showed generally an increasing trend with leaf age (Fig. 73b). A similar pattern was found for ϵ' (Fig. 73c). The general increase in ϵ' with leaf age is consistent with the view that cell walls lose their elasticity as leaves become older.

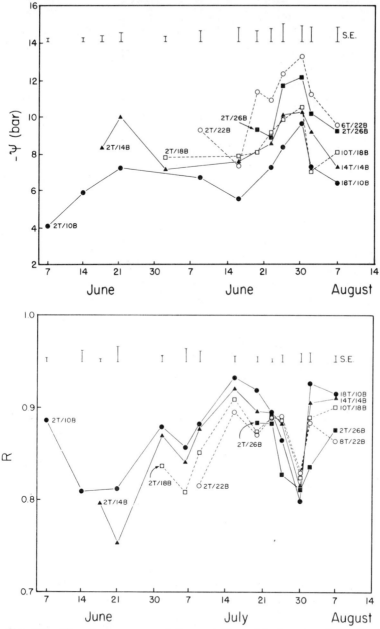

Fig. 72. The time course of the variables of leaf water status for leaves at different positions on the plant. Each leaf is designated by two numbers; the first shows the position from the top (T), the second the position from the bottom (B), of the plant. (a) Water potential (Ψ). (b) Relative water content (R) (after Kassam and Elston 1976).

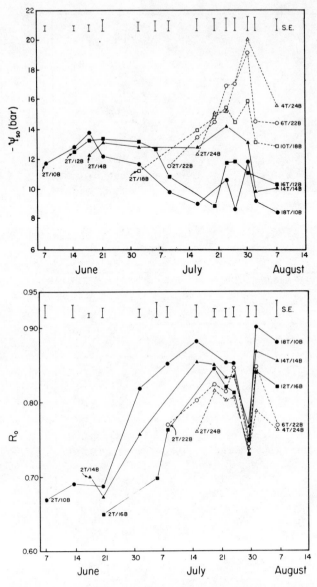

Fig. 73. The time course of some parameters of water status for leaves at different positions on the plant. Leaf designation is the same as in Fig. 72. (a) Solute potential at zero turgor (ψ_{so}). (b) Relative water content at zero turgor (R_o). (c) Slope of the linear regression between pressure potential and relative water content (ϵ') (after Kassam and Elston 1976).

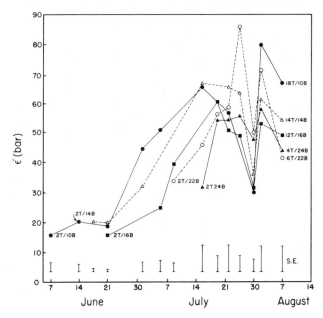

Fig. 73. (*Continued.*)

Effects of the Soil Water Status and Aerial Environment   The effect of soil water stress on the internal water balance of field beans was studied in polythene-covered plots in the field in a sandy loam soil under different irrigation treatments (Elston et al. 1976; Karamanos 1978b). There were three irrigation treatments, wet (*W*), medium (*M*), and dry (*D*), differing in the length of the irrigation cycles. Thus the plots of the *D*-treatment were subjected to a single irrigation cycle, those of the *M*-treatment to 2 to 3 cycles, while those of the *W*-treatment to about 12 cycles (they were irrigated about twice a week). In this way the water potential at dawn of the plants in the *W*-treatment never fell below −3 bar, while the corresponding values in the *M*- and *D*-treatments were −5 and −8 bar. The effect of the aerial environment was examined by studying the responses of the variables and parameters of plant water status in two contrasting growing seasons as well as within each season. Two field experiments were made, one in 1974 and the other in 1975. In 1974 the season was wetter and cooler than in 1975.

The course of Ψ in the three treatments during the two seasons is shown in Fig. 74. A similar pattern was followed by *R*. In both

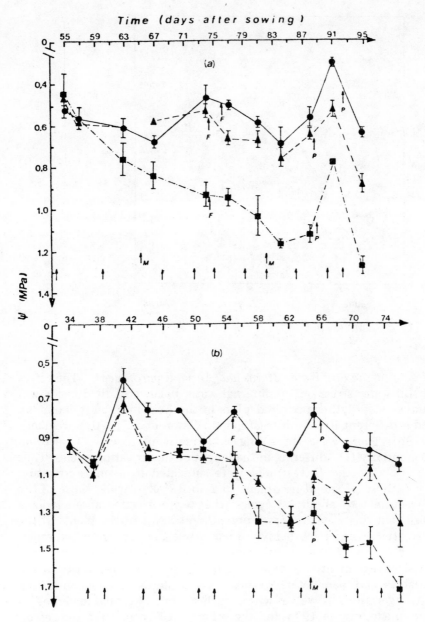

Fig. 74. The time course of the water potential (Ψ) of the youngest fully expanded leaf in three irrigation treatments (wet: ●——●, medium: ▲----▲, dry: ■–·–·–■) and two growing seasons. (a) Wet season (1974). (b) Dry season (1975). Arrows indicate the irrigation timing in the wet and medium (M) treatments. The beginnings of flowering (F) and podding (P) are also designated (after Karamanos 1978b).

years, the Ψ in the W-treatment was kept relatively constant at values consistently higher than those in the D-treatment, where Ψ followed a generally falling trend. In the M-treatment, Ψ took intermediate values between the two extremes. In all treatments the values of Ψ were significantly more negative in the drier year (1975).

In both years ψ_{so} did not show any consistent pattern in the W-treatments (Fig. 75). However, in the D-treatments ψ_{so} showed a falling trend during the second half of the experimental seasons, reaching values much lower than those in the W-treatments. The irrigations in the M-treatments restored the already fallen ψ_{so} to higher values, but with a lag of some days. As in the case of Ψ, the values of ψ_{so} were lower in the drier years, especially in the M- and D-treatments. The falling trend of ψ_{so} in the D-treatments can be associated with the increasing soil dryness, taking into account that these treatments were not irrigated during the observations. The atmospheric dryness in 1975, which brought about a more intense depletion of soil water by means of both plant roots and evaporation, is probably responsible for the lower values of ψ_{so} recorded in that year.

The pattern of R_o with time was quite distinct between the extreme treatments in 1974, the values for the D-treatment being significantly lower than those in the W during most of the season (Fig. 76). The same did not happen in the drier year, where the time courses of R_o in the different treatments were overlapping. In both years, no consistent trend was detectable in any treatment. Furthermore the response of R_o to irrigation was much quicker than that of ψ_{so}. The course of ϵ' in the wet year was similar to that of R_o (Fig. 77), while no detectable differences could be traced in 1975. The similarity of the patterns of both R_o and ϵ' suggests that cell-wall elasticity mainly influenced R_o (Karamanos 1978b). Moreover, the approximately parallel patterns of both R_o and ϵ' for all treatments in 1974 imply that these parameters were also influenced by short-term environmental factors.

In conclusion, it is possible to distinguish two mechanisms that control internal water status in field beans. The first operates through solute potential adjustment and is a result of solute accumulation in the tissues, as the consistent drop in ψ_{so} under dry conditions suggests. That this drop in ψ_{so} is brought about mainly by an osmotic adjustment is shown by the fact that ψ_{so} is well correlated ($r > 0.85$) with the osmotically active component of the dry weight of the tissue (Kassam and Elston 1976). This lowering of ψ_{so} follows the progressive depletion of the soil water and is not easily reversible to higher values after rewatering. Hence it constitutes an adaptive

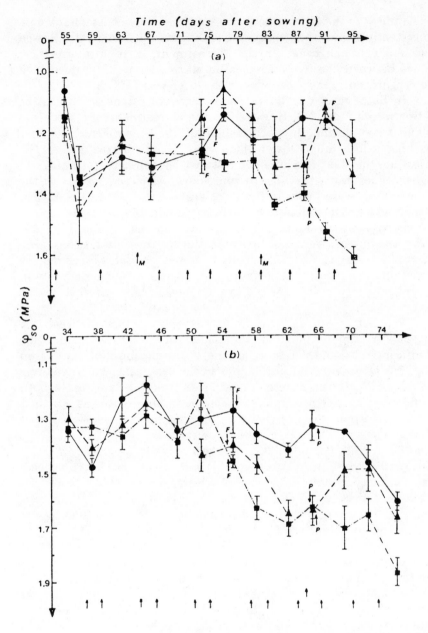

Fig. 75. The time course of the solute potential at zero turgor (ψ_{so}) of the youngest fully expanded leaf in three irrigation treatments and two growing seasons. (a) Wet season (1974). (b) Dry season (1975). Symbols the same as in Fig. 74 (after Karamanos 1978b).

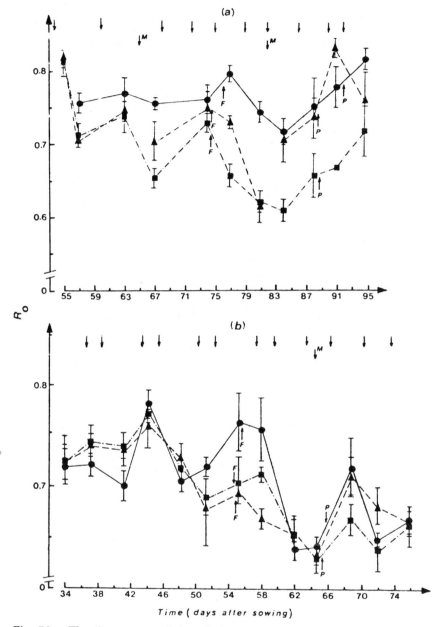

Fig. 76. The time course of the relative water content at zero turgor (R_o) of the youngest fully expanded leaf in three irrigation treatments and two growing seasons. (a) Wet season (1974). (b) Dry season (1975). Symbols the same as in Fig. 74 (after Karamanos 1978b).

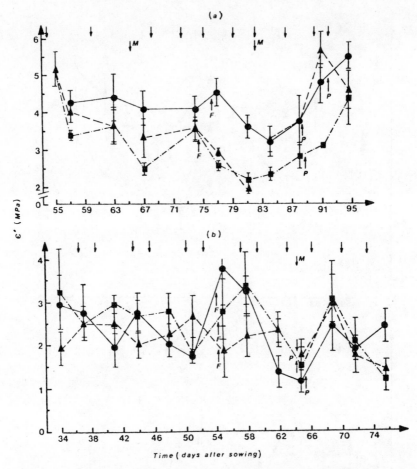

Fig. 77. The time course of the slope of the linear regression between the pressure potential and relative water content (ϵ') of the youngest fully expanded leaf in three irrigation treatments and two growing seasons. (a) Wet season (1974). (b) Dry season (1975). Symbols the same as in Fig. 74 (after Karamanos 1978b).

mechanism in field beans. The second mechanism operates through pressure potential adjustment, which results mainly from fluctuations in R_o and ϵ'. The patterns of these parameters indicate that they are not as sensitive to soil water stress as ψ_{so}. When plant tissues were not seriously water stressed, as in 1974, they tended to lower ϵ' (increase wall elasticity) with progressively increasing drought. A concomitant drop in R_o is also evident. In this way cells maintain turgor as high as possible for a given water loss. This behavior of ϵ' develops even at low soil water stress (treatments are separated from the beginning of the observations and not from the middle of the season, as ψ_{so}; Fig. 77) and is easily reversible on the removal of stress by irrigation. Moreover, ϵ' follows the day-to-day fluctuations in the evaporative demand (Karamanos 1976). On the other hand, the lack of treatment differentiation in ϵ' in the dry year, together with the much lower values of ϵ' for all treatments in 1975, suggest that the lowest limit of ϵ' was reached in that year, even in the W-treatment. Thus no separation of the treatments was possible, and ϵ' responded only to the short-term fluctuations of the aerial environment.

The importance of the osmotic adjustment for the water relations of field beans can also be demonstrated in Fig. 78. A more sudden drop in Ψ for a given drop in R was observed in 1975. Such behavior enables the tissues to lower their water potential, so that they can sustain the driving force necessary for water uptake. However, on no account should this occur by means of a drop in cell turgor which is indispensable for many physiological processes (Hsiao 1973). By comparing the slopes of both ψ_p and ψ_s against R in Fig. 78, it is evident that the slope of the regression between ψ_s and R was much steeper ($p < 0.001$) in the dry year, while no difference was detectable between the slopes of ψ_p against R in the two years. Thus the desirable drop in Ψ is caused mainly by means of osmotic adjustment in species such as *Faba* beans. This contrasts with adjustment in wall elasticity in xeric grasses (Maxwell and Redmann 1974).

Water Loss

Approach The transpiration of field beans under field conditions is considered in this section. Transpiration was evaluated indirectly using the soil water balance method. According to this method, the rate of water depletion in the root zone of a crop during a given time interval is taken as equivalent to the water consumption by this crop, provided that drainage does not occur (Tanner 1968). In view of this restriction, results only from the dry and moderately wet plots will be presented, for the frequent irrigations in the W-treatments were necessarily accompanied by groundwater drainage. The rates of soil water depletion were determined from successive readings of nylon–stainless steel resistance units (Farbrother 1957) buried at

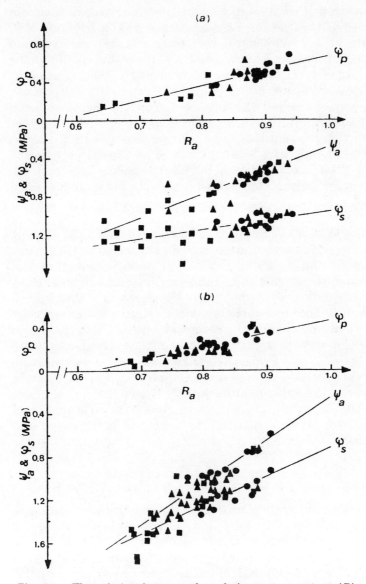

Fig. 78. The relation between the relative water content (R) and the water (Ψ), pressure (ψ_p), and solute (ψ_s) potentials in two growing seasons. (a) Wet season (1974). (b) Dry season (1975). Results from three irrigation treatments (wet: circles; medium: triangles; dry: squares). The lines represent the fitted linear regressions. Note the steeper slopes of the regression lines between R and ψ_s in the drier year (after Karamanos 1978b).

4 different depths (20, 40, 60, and 90 cm). These units also made possible the determination of the soil matric potential, which will be taken as identical to the soil water potential (Ψ_{soil}). The depth of 90 cm was taken as representing the lowest limit of the root zone of this crop (Kutschera 1960). Transpiration rates were expressed per unit leaf area of the crop.

Unfortunately, no measurements of stomatal resistance were taken in the field. Thus speculations on the stomatal behavior of field beans will be made on the basis of results on broad beans taken in a growth cabinet (Kassam 1973).

Transpiration and Stomatal Behavior Crop transpiration rate (E) decreased with a drop in either plant (Ψ) or the average soil water potential (Ψ_{soil}) (Fig. 79).

The drop in E was more abrupt with Ψ_{soil} than with Ψ. Ψ_{soil} affects transpiration rate indirectly through its effects on the uptake of soil water. A small drop in Ψ_{soil} is accompanied by a much greater increase in the soil resistance to water movement. Such a decrease in soil water uptake may affect plant water status by activating the mechanisms that control transpiration. There is a more direct association between E and Ψ. The reduction in E with falling Ψ implies that the resistance controlling water loss increased with the drop in Ψ. It was found (Kassam 1973) that at saturating light intensities, the stomata of broad beans closed in the range of Ψ between −8 and −10 bar (Fig. 80).

Given that the values of Ψ in the field experiment were below −9 bar, one could ascribe the drop in transpiration to an increase in the stomatal resistance because of the low Ψ. Nevertheless, the gradual fall in E even at Ψ far below −12 bar suggests that the stomatal behavior shown in Fig. 80 does not hold under field conditions. This may be due to the preconditioning of the plants to the drought stress during the progressive soil drying (Brown et al. 1976).

Kassam (1973) observed a different behavior between the stomata of the two surfaces of broad bean leaves. At a given light intensity, the resistance of the lower epidermis was always lower than that of the upper epidermis, while the saturating light intensity was about 10 times higher for the upper leaf surface. This difference was attributed to wider opening of the stomata of the lower surface rather than to differences in stomatal densities between the two surfaces. At light intensities below saturation, the increase in stomatal resistance with the lowering of leaf water potential is less abrupt than that occurring at and above saturating light intensities (Fig. 81).

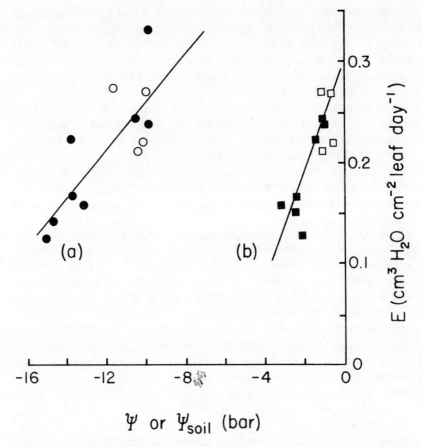

Fig. 79. Relationship between crop transpiration rate (E) and leaf (Ψ) and soil (ψ_{soil}) water potentials. Circles: E vs. Ψ; squares: E vs. Ψ_{soil}. Results from two treatments, dry (filled symbols) and moderately wet (open symbols). The lines (a) and (b) represent the fitted linear regressions. (a) $Y = 0.50 + 0.24X$ ($r = 0.82$). (b) $Y = 0.30 + 0.54X$ ($r = 0.81$) (after Karamanos 1980).

Water Uptake

Approach The extension and activity of the root system determine the possibilities of the plant to absorb water. All relevant information about *V. faba* from both field and laboratory experiments will be presented here.

Information on the growth in depth of the root system under field conditions will be taken indirectly from the measurements of the nylon-stainless steel resistance units (Karamanos 1980). The depletion of water at a given soil layer will be taken as an indication of the presence of roots in that layer. The activity of the root system in the various soil layers was assessed by studying the resistances to water flow in these layers in a drying soil. The calculation of the resistances was based on the approach of van den Honert for the soil–plant system (Karamanos 1980).

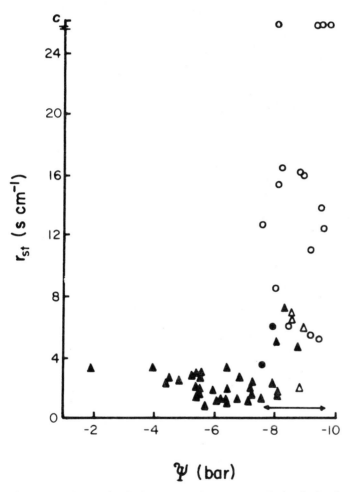

Fig. 80. Relation of the stomatal resistance (r_{st}) of the lower leaf surface and leaf water potential (Ψ) for broad beans. Filled symbols: not flaccid leaves; open symbols: flaccid leaves; triangles: open stomata on the upper surface; circles: closed stomata on the upper surface. C indicates no measurable diffusion of water vapor. The double arrow shows the range of Ψ in which the stomata of the lower surface were closed or closing (after Kassam 1973).

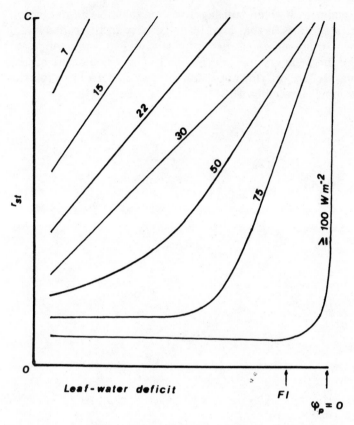

Fig. 81. Proposed relation between stomatal resistance (r_{st}) and leaf water deficit for the upper surface of broad bean leaves at different irradiances shown on the curves. The saturating irradiance is taken as equal to 100 W m^{-2}. Arrows indicate when leaves become flaccid (*Fl*) or their turgor drops to zero ($\psi_p = 0$) (after Kassam 1973).

Root Extension   It is generally accepted that increasing soil water stress in the surface soil layers induces compensatory root extension to deeper, unexploited soil layers (Newman 1966; Klepper et al. 1973). El Nadi et al. (1969), working with broad beans growing in deep containers, found that progressive soil drying promoted deeper root growth, although the total dry weight of the root system was the same under wet and dry soil conditions. This happened because root extension in the deeper layers occurred at the expense of formation of new roots in the layers close to the soil surface (Fig. 82).

In general, conditions of high soil water content (at or even above field capacity) are extremely favorable for maximum root development of broad beans (Jones 1963).

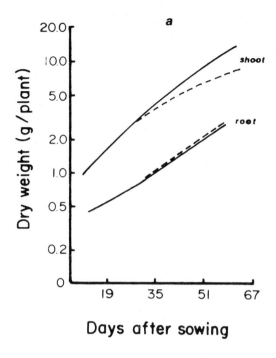

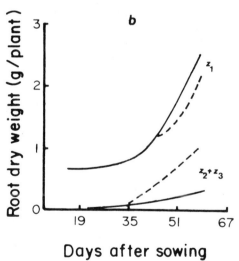

Fig. 82. (a) The growth in dry weight of shoots and roots of broad bean plants grown at 17°C in wet (continuous line) and dry (broken line) soil. (b) The growth in dry weight of roots in the upper (0 to 20 cm, z_1) and lower (20 to 60 cm, $z_2 + z_3$) layers of soil for broad beans grown at 17°C in wet (continuous line) and dry (broken line) soil (after El Nadi et al. 1969).

Drought was shown to induce deeper root extension also under field conditions. This is shown in Fig. 83 for two field bean crops, irrigated and dry. Increasing soil dryness promoted a quicker exploitation of the deeper soil layers.

Roots of the unirrigated plants reached the depth of 90 cm about 72 days after sowing, while the normal maximum rooting depth for some species of the genus *Vicia* seldom exceeds 70 to 80 cm in light soils (Kutschera 1960). It is worth noting that root growth in the drought-stressed plants proceeded beyond the pod-forming stage, which is regarded as the point of cessation of root growth in field beans (El Nadi et al. 1969; Sprent et al. 1977).

Root Activity The effectiveness of the root system to absorb water in a given soil layer is a function not only of the root mass in that layer, but also of the average root age. It is generally accepted (Scott Russell 1977) that water can enter most readily into the young unsuberized zones of roots, but absorption is not necessarily confined to these zones. In general, the rate of water uptake is greater near the root apex than at increasing distances away from it. However, experiments with broad beans (Brouwer 1953) showed that the extent to which water enters different parts of young roots is modified by the rate of water absorption. Thus the major contribution of the portion near the root tip at low rates of water uptake decreased in favor of the more remote portions with increasing uptake rates. Other external factors may also play a role in the entry of water into different parts of young roots (Newman 1974).

The activity of the root system of field beans under field conditions was assessed from the rates of soil water depletion at different depths (Karamanos 1980). Vertical (upward) soil water movement, which operates in parallel to the uptake of water by roots in determining the water depletion at a given layer, was neglected. Such a movement was not considered as introducing significant errors in the calculations for a drying soil, except for the surface layer where direct evaporation plays an important role. It can be seen that the maximum rate of water depletion was shifted toward deeper soil layers as the soil dried out (Fig. 84).

Such a shifting can be ascribed (1) to a decrease in soil hydraulic conductivity which greatly impairs soil water movement toward roots, (2) to the presence of younger roots in the deeper (younger) portions of the root system, and (3) to the death of roots by desiccation in the upper soil layers. Taylor and Klepper (1971) observed a production of new roots in the dry soil layers after irrigation in

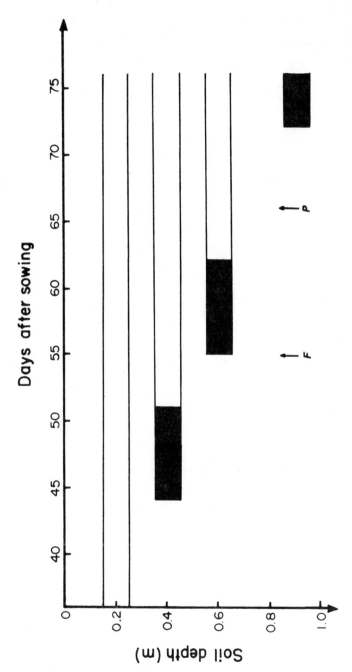

Fig. 83. The presence of roots at different soil depths as a function of crop age for two field bean crops growing in the field in 1975 [☐: wet crop (watered twice a week); ■: dry crop]. The first decrease in soil water detected by means of resistance units at a given depth was taken as indicating the presence of roots at that depth. Arrows indicate the beginnings of flowering (*F*) and podding (*P*) (Karamanos unpublished).

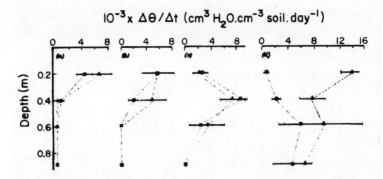

Fig. 84. Profiles of the rates of soil water depletion ($\Delta\Theta/\Delta t$) in four selected time intervals during the development of two field bean crops, moderately wet (▲) and dry (■). (a) 37 to 41 days, (b) 48 to 51 days, (c) 58 to 62 days, and (d) 72 to 76 days. The moderately wet crop was watered on day 62 (after Karamanos 1980).

cotton plants. This may explain, in parallel with the increased removal of water by evaporation, the high rates of soil water depletion at the depths of 20 and 40 cm after an irrigation (Fig. 84d). The calculation of the resistance to water uptake at separate soil layers revealed that the resistances increased with soil depth for a given value of the soil water potential (Fig. 85).

Once possible effects of soil structure are eliminated, such behavior is attributed either to a significant axial resistance to water flow along the root xylem and/or to an increase in the radial root resistance with soil depth (Karamanos 1980). The latter may be associated with changes in root ultrastructure brought about by increasing soil compaction and mechanical impedance to root extension (Wilson et al. 1978). Possible effects of anaerobiosis may also be bound up with the increased resistance to water uptake at deeper layers. It appears therefore that it is important to avoid an extreme desiccation of the surface soil layers which ensure an easier supply of soil water to plants.

Drought-Stress Effects

Plant water stress reduces both the dry matter (Karamanos 1976) and the seed yield (El Nadi 1969; Karamanos 1976). This reduction in final yield arises from water-stress effects on many physiological processes and yield determinants. The research progress on the effects of drought stress on leaf-area growth and nodulation is reported below. Both of the latter processes are extremely important, since the former determines the capacity of the photosynthetic system while the latter controls the nitrogen nutrition of these plants.

Effects of Drought Stress on Leaf-Area Growth

Drought stress was found to control effectively the total active leaf area in field beans, as it does in many other plant species (Karamanos 1978a). Results from two years of field experimentation under three different irrigation treatments showed a high positive linear relationship between the final leaf area and the average plant water potential during the growing season (Fig. 86).

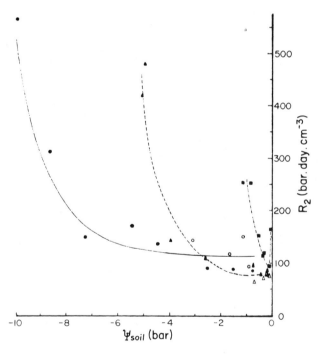

Fig. 85. The resistance to water flow at three soil depths (R_z) as a function of the soil water potential (Ψ_{soil}) at the corresponding depths. ●,○: 20 cm; ▲,△: 40 cm; ■,□: 60 cm. The curves for each depth were drawn by eye (after Karamanos 1980).

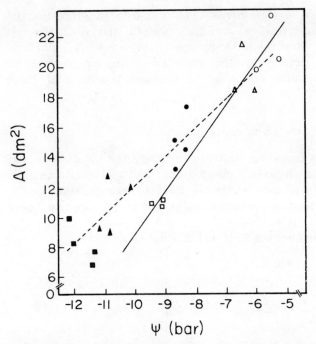

Fig. 86. The relationship between the total leaf area of field-grown field bean plants (A) and the average plant water potential prevailing throughout the growing season (Ψ). Results from a wet (1974, open symbols) and a dry (1975, filled symbols) growing season. There were three irrigation treatments in each season, wet (circles), medium (triangles), and dry (squares). The lines represent the fitted linear regressions for each year. Continuous line: 1974 ($Y = 37.3 + 28.4X$, $r = 0.96$); broken line: 1975 ($Y = 31.4 + 19.4X$, $r = 0.86$) (after Karamanos 1978a).

The number and size of the living leaves determine the total leaf area per plant at a particular moment. The number of living leaves at any time is the difference between the total number of leaves produced up to that time and the number of dead leaves. Thus to understand how water stress affects leaf-area growth, it is important to assess the effects of water stress on leaf production, leaf death, and leaf expansion.

Drought Stress and Leaf Production   The leaves of *V. faba* unfold when each lamina attains a significant proportion (0.25 to 0.65, depending on leaf position and external factors) of its final size. Thus the production of photosynthetically active leaves depends on two mechanisms quite different in their physiology, namely, leaf appearance and unfolding. The sensitivity of both mechanisms to water stress was significantly affected by plant development. The rates of leaf appearance and unfolding responded negatively to water shortage only after flowering, although plants experienced serious drought stress long before the beginning of flowering (Karamanos 1978a). Plant water potential exerted a cumulative long-term effect on both mechanisms with the rate of leaf appearance being slightly more sensitive than that of unfolding to water stress. The growing season also affected the critical values of cumulative plant water potential below which the rates of both mechanisms started falling. In the wet year (1974) the critical plant water potential was about −8 bar, while in the drier year (1975) the corresponding value was about −10 bar. This difference may be attributed to a development of adaptive responses of the plants in the dry year.

Drought Stress and Leaf Death Leaf death of the older leaves was hastened by drought stress in field beans (Finch-Savage and Elston 1976). In comparison with leaf appearance and unfolding, the rate of leaf death was more sensitive to water stress (Karamanos 1978a). It is suggested (Kozlowski 1976) that water stress alters the hormonal balance of the leaves, thus promoting both an overall senescence and the hydrolytic reactions in the abscission layer of the petioles. Brady et al. (1974) attributed the drought-induced senescence to a change in the ABA/cytokinin balance. Indeed, water stress promoted the formation of growth inhibitors (phenolics and ABA) at the expense of growth stimulators in field beans (Pustovoitova 1972).

Drought Stress and Leaf Expansion To assess the effects of external factors on leaf expansion, a quantitative expression of the process is necessary. The procedure followed with field beans consists of two stages:

1. Fitting of mathematical curves to the experimental data relating laminar expansion to time. Since the pattern of leaf growth against time is sigmoid, the asymptotic function proposed by Richards (1959) was used. This function is quite flexible and can be adapted to a wide range of forms of sigmoid curves.

2. Use of the constants and parameters of the fitted function to derive other biologically meaningful parameters (Dennett, Auld, and Elston 1978). The parameters used for the description of

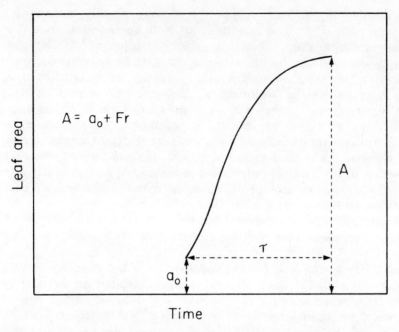

Fig. 87. The parametrization of the function used for the description of laminar expansion in field bean leaves. The final area of a leaf (A) is the sum of the area at leaf unfolding (a_o) plus the product of the mean growth rate during expansion (F) and the duration of growth (τ) (after Karamanos 1979).

leaf expansion in field beans are: the area at which leaves unfold (a_o), the duration of expansion (τ), the mean absolute growth rate (F), and the final area (A) (Fig. 87).

In this way, the final leaf area is derived from the other growth parameters as follows:

$$A = a_o + F\tau \tag{51}$$

The final area was considerably reduced by drought stress. This reduction in A was brought about by a drop mainly of the expansion rate (F) and, in second place, of the area at unfolding (a_o). No consistent effect on the duration of growth was detected in the two years of experimentation. The dependence of the various growth parameters on water stress can be evaluated by examining the linear relations between any of them and the average plant water potential prevailing during or before leaf expansion (for the area at unfolding) (Fig. 88).

The expansion of the leaf lamina is the result of two processes, cell division and enlargement. Cell division dominates during the early stages of leaf expansion, while cell enlargement dominates in

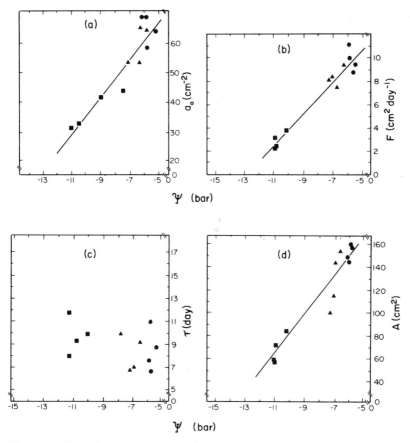

Fig. 88. The relation between each of the parameters used for the description of laminar expansion and the average leaf water potential prevailing during expansion (the area at unfolding was correlated with the average Ψ for 6 days before unfolding). Results from three irrigation treatments and four different leaf positions (pooled together) in 1974. Symbols the same as in Fig. 86. The fitted linear regressions are also shown, except for (c) where no relation was apparent. (a) Area at unfolding (a_o), $Y = 100.7 + 64.5X$ $(r = 0.94)$. (b) Mean absolute growth rate (F), $Y = 17.8 + 14.2X$ $(r = 0.98)$. (c) Duration of growth (τ). (d) Final area (A), $Y = 247.2 + 168.3X$ $(r = 0.96)$ (after Karamanos 1979).

the later stages. In field beans both processes were found to proceed simultaneously until about 50 to 60% of the final area is attained (Farah unpublished). Beyond this stage, cell enlargement determines leaf expansion. Both processes were reduced by drought stress to the same extent by means of a reduction in their rates and not in their duration (Farah *ibid*.). This may also explain why the rate of overall leaf expansion was unaffected by water stress.

In comparison with leaf production and death, laminar expansion was the most sensitive to drought stress (Karamanos 1978a). Thus the differences in total leaf area between well-watered and drought-stressed field beans is mainly caused by the smaller leaf size of the stressed plants and secondarily by a fall in leaf production or an increase in the rate of leaf death. Accordingly, the order of decreasing sensitivity to water stress of the mechanisms determining leaf-area growth in field beans is the following: leaf expansion, leaf death, and, finally, leaf production.

Effects on Nodulation

The formation of nodules in the root system of legumes is of paramount importance for their nitrogen nutrition. Under conditions of good nodulation, field beans do not respond to N-fertilizers and thus are regarded as self-sufficient for nitrogen (McEwen 1970; Sprent and Bradford 1977). Among the various external factors influencing nodulation (soil and air temperature, irradiance), drought stress was found to exercise a decisive impact. The results presented below come mainly from field experiments carried out at Dundee, Scotland, by Dr. J. I. Sprent and her coworkers.

The nitrogen-fixing activity (expressed as acetylene-reducing activity) of the nodules in field beans under field conditions was related in a curvilinear fashion to soil water content (Fig. 89).

The maximum activity was observed at about field capacity. Activity fell steadily as the soil became drier and also declined when the water content exceeded field capacity, probably because of the lack of oxygen. A significant fall in N-fixation coincided with the onset of visible wilting in the lower leaves. The close association between nodule activity and soil water is also shown by the fact that restoration of the activity to its maximum values was observed when the dry plots were rewatered.

The reduction in the nitrogen-fixing activity of the drought-stressed plants was associated primarily with the number of developed nodules. The normal ontogenetic increase in nodule number per

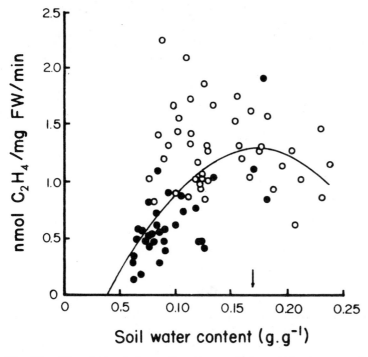

Fig. 89. The relation between the nitrogen-fixing activity (expressed as acetylene reduction) and the gravimetric soil water content for field-grown field bean plants. ○: means of three plants, ●: means of six plants. The arrow indicates the soil water content at field capacity for the given soil (after Sprent 1972b).

plant was enhanced when plants were transferred to a treatment with more water, but depressed when their water supply was reduced (Sprent et al. 1978) (Fig. 90).

During periods of water stress, it is quite likely that plants are unable to develop all nodules previously initiated. Furthermore, water supply increased the size of the nodules, but without any effect on nodule weight and specific activity (Gallacher and Sprent 1978).

It has not yet been fully established how the water status of the nodules is affected by the water relations of the rest of the plant. It appears that nodules are supplied with water through the vascular connections with the root rather than from their surface (Sprent 1972b). The surface of the nodules is adapted for gaseous exchange and, accordingly, is more adapted to water loss than uptake (Sprent

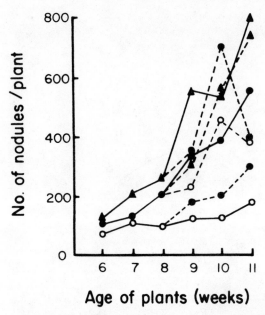

Age of plants (weeks)

Fig. 90. The course of the number of nodules per plant with crop development for plants growing under different water regimes. ○: water stress, ●: sufficient water, ▲: excess water. ———: plants exposed throughout to one treatment, -----: plants transferred from one water treatment to another, as indicated by symbols (after Sprent et al. 1978).

1972a). Thus nodules are desiccated when bean plants show symptoms of water stress (for example, wilting of the lower leaves). Under these conditions, water is removed from the nodules and is transferred through the vascular system to the more seriously stressed overground plant parts.

5

G. M. SIMPSON

The Value of Physiological Knowledge of Water Stress in Plants

IN PLANT BREEDING

The Basic Problem

Plant breeding relies heavily on the identification of superior traits in individuals. These individuals are either removed from the general population and multiplied to constitute a new population with superior attributes or crossed to other plants to combine the specific characteristics that are desirable from an agronomic or utilization perspective (Eslick and Hockett 1974).

The central practical problem for plant breeders using any one of the common approaches such as selection, crossing, back-crossing. hybridization, and so on is the final physical technique of sieving out the one, dozen, or fifty plants with desirable traits from the tens or even hundreds of thousands of individuals that constitute the population (Simpson et al. 1979). In the case of increasing crop yield, great advances have been made in the course of this century simply through the application of the gravimetric balance to selection for dry weight (Borlaug 1968). The second most important selection tool has probably been the artificial disease epidemic for the selection of plant disease resistance. Both of these advances have occurred because it has been possible to separate discrete classes, or individuals, by the application of simple but powerful sieving techniques.

At first sight it would seem possible to recover plants with superior drought resistance from a large population simply by the application of drought under either natural or artificial conditions. However, resistance to drought in plants is not a simple trait but rather is a complex of traits which include characteristics of both avoidance and tolerance (Hurd 1976; Levitt 1972). Avoidance can be variously achieved by such attributes as increased root growth, early maturation, premature reduction in leaf area by leaf senescence, or reduced evapotranspiration through leaf-rolling and stomatal closure (Stout and Simpson 1978). On the other hand, tolerance of drought stress may be affected by osmoregulation, differentiation, and development of plasmatic resistance (Kaul and Crowle 1971; Stout et al. 1978). Resistance to drought stress, as with cold stress (Burke et al. 1976) and salinity stress (Flowers et al. 1977), thus depends on a complex of attributes in the plant that confers both survival and a range of productivity potential at various stages of the life cycle.

From a practical point of view the application of drought stress, in a measured and repeatable manner, is very difficult at the scale of the very high plant populations normally used for plant breeding. Numerous techniques have been tried, ranging from simple pot experiments where water is withheld (Perroux 1979; Quarrie and Jones 1979) to extensive irrigation experiments in an arid or semiarid environment where water is applied or withheld at various stages of the life cycle of the plant (Blum 1974; Kaul 1974). In the main, often despite considerable monitoring of physiological and growth parameters, these approaches have merely uncovered high- or low-yielding plants because the final selection criterion was a simple index such as grain yield at harvest or forage yield at haying time. Such approaches have generally failed to disclose those individual genetic traits affecting drought resistance which could be selectively removed for further recombination to produce, for example, a plant that is specifically resistant to drought stress at anthesis, or a plant with avoidance of drought stress at grain-filling combined with tolerance to a short stress period of a particular magnitude at any specific stage of its life cycle.

Plant breeding for any form of drought resistance has been apparently hampered by at least three requirements:

1. Identification of specific parameters that represent the significant traits that collectively confer survival and productivity for each stage of the life cycle.

2. Development of a simple practical sieve for selecting out individual plants possessing each of the above parameters.
3. The development of routine, large-scale methodology for creating defined and repeatable water stress in the environment of large populations of either field-grown or laboratory-grown plants.

To date, the physiological knowledge about the response of plants to water stress has not been widely used by plant breeders. In part this can be attributed to the simple fact that breeders have been faced with long and expensive selection programs, using very large populations, where for economic reasons yield has been the most important characteristic to identify. The introduction into breeding programs of less well-defined characteristics, with the exception of plant disease resistance and some quality characteristics, is not likely until at least the following conditions have been met:

1. The physiological response (*PR*) within the plant to a specific water stress at a specific stage of the life cycle has been accurately described.
2. A simple physical parameter (*P*) in the plant, easily monitored, that correlates highly with the observed *PR* has been identified.
3. Proof that sufficient genetic variability exists within a population for the specific *P* to warrant a plant selection program.
4. Establishment of a specific water stress that simulates the ultimate field conditions for crop growth under large-scale laboratory or field conditions to elicit *P*.
5. Demonstration that selection for *PR* and its recombination with other characters does not have negative effects on such prime characters as yield, disease resistance, or quality of vegetative and seed products.

Further consideration of each of these conditions is worthwhile. The problem of first accurately describing the *PR* and then finding a single simple *P* that integrates the water stress experienced by the plant is illustrated by reference to an early paper by Fischer (1973).

The experiments indicated that plants exposed to a soil drying cycle for a given number of days under carefully controlled conditions reached different levels of plant water stress depending on their stage of development. In addition, there were always large differences in water potential between the top of the plant and the bulk soil. This was taken to indicate that soil water potential may not be a good estimator of plant water potential. Nevertheless the water stress measured within the plant was associated in a consistent manner with

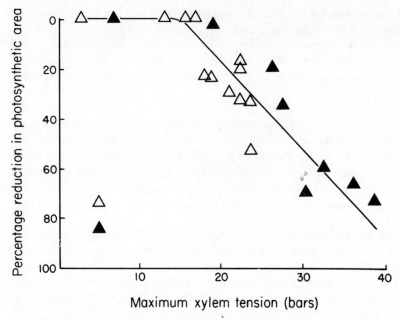

Fig. 91. Effect of degree of water stress in wheat on the reduction in photosynthetic area per shoot during the stress cycle, expressed as a percentage of the photosynthetic area present immediately prior to stress. △: stress measured 2 days before ear emergence; ▲: stress measured after this date (after Fischer 1973).

the reduction in photosynthetic area and sink size (grain number), both of which are major yield-determining parameters (Figs. 91, 92). However the difficulty in applying this knowledge—for example, to plant selection—was summed up by the author in the following manner.

> In applying the above information on plant responses to water stress to the field situation, one is particularly limited by the problem of an appropriate single parameter which adequately describes the water stress experienced by the plant during a drying cycle. As well as the maximum degree of stress, which formed the stress indicator in this study, there is evidence that the duration of stress may be important. Also the part of the plant on which plant water stress is measured should be related to the plant response with which one is concerned. In this respect xylem tension measured on the stem may provide a more useful indicator of plant water stress than relative water content of leaves, especially when stress effects on the developing ear may dominate the overall yield response.

Faced with this kind of uncertainty of interpretation, a plant breeder would be reluctant to opt for using either relative water content or xylem tension as physical parameters to monitor because of the obvious complexity in interpretation. Nevertheless, in this paper Fischer clearly demonstrated the first step, namely, that of accurately describing a significant physiological response to water stress: in this case, the quantitative relationship between a major component of yield (grain number) and water stress applied during the most critical stage of grain development (Fig. 92). A great number of published papers ranging from irrigation experiments under field conditions to elaborate experiments in controlled environments testify that drought stress can affect any stage of development of a plant to bring about a wide variety of modifications of subsequent growth and development.

The second step, to identify a single, easily measured parameter in the plant that directly reflects the interaction between the external water stress and a significant facet of growth, has been fraught with

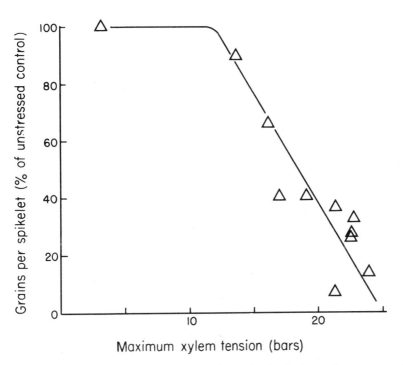

Fig. 92. Reduction in spikelet number of wheat stressed over a period of 20 days commencing 15 days before ear emergence (after Fischer 1973).

difficulty. Such potential indices as relative saturation deficit (Dedio 1975), tissue osmotic potential (Stout and Simpson 1978), leaf water potential (Gardner 1973), stomatal behavior (Jones 1979), leaf and stem growth (Hsiao 1973), photosynthetic activity (Kaul 1974), membrane permeability (Bewley 1979), and plant hormones (Quarrie and Jones 1979; Simpson et al. 1979) have been investigated. None of these parameters alone has a simple relationship to plant yield, particularly of grain and fruits. Nevertheless, in most cases there is some degree of relationship to growth.

The two key problems that have hindered the effective use of any one of these parameters are (1) the lack of correlation of a single index with such a complexly determined character as yield which in effect is the resultant of the complete ontogeny of the plant shaped by all the variables that have influenced it from germination, and (2) the development of a simple test that is both easily applicable to a single plant yet suited for screening many thousands of plants quickly and economically.

Kaul and Crowle (1974) have suggested using a single index derived mathematically from several parameters, each relatively easily measured in single plants. Their index adequately reflected the general drought-resistance and yield characteristics of the limited number of wheat varieties investigated. This was based on their known field performance over many years in a range of annual environments in a semiarid climate. Nevertheless, their index could not, in the final analysis, be used under general field plot conditions. They found that in field plots the variation in individual plant response at individual microsites was so great, within an apparently uniform inbred line of wheat, that it could obscure differences between genotypes. This work thus points strongly to (1) the need for using a very uniform soil or hydroponic environment when water stress is used as a screening procedure for comparing individual plants, and (2) the need for proof, before beginning a screening procedure with self-pollinated species, that genotypes selected over 15 or 20 generations for such characteristics as yield or disease resistance are indeed also uniform for their water relations characteristics.

Application of Physiological Knowledge to Plant Breeding

A glance at some of the published work of the last several years indicates that examination of the genetic basis of the responses within plants to water stress is not a high priority on the part of plant physiologists when they study the basic physiological responses of plants to water stress (Table 22). While Table 22 is not a comprehensive list, it is a good cross-section of interest at this time, and some general conclusions can be drawn from it. Clearly there is a wide range of species, mostly economically important crop, horticultural, and forest species being examined for their response to water stress. In the great majority of cases only one cultivar within the species was investigated. In a majority of cases the focus of the study has been on one and, very occasionally, two or more plant responses to water stress. The single-factor approach makes the task of assessing the integrated response of the plant to water stress difficult. No doubt this also reflects the increased technical difficulties associated with measuring several parameters simultaneously. It must also reflect the complexity, tediousness, and cost of carrying out many of the analyses quoted. Of the wide array of plant characteristics measured, only a few are relatively simple to measure, either in physical or economic terms, over a short period on a considerable number of plants. Of the examples cited probably the easiest measurements would be fatty acid content, epicuticular wax formation, pigment changes, leaf number and area, sugar content, leaf extension rate, tillering and internode number and length, and soil CO_2 production. Perhaps the inference might be drawn that physiologists, by choosing complicated procedures, demonstrate they are interested primarily in understanding the physiology of the response and only secondarily interested in collaborating with plant breeders to improve the possibilities of plant breeding for drought resistance.

Another important conclusion can be made from a detailed study of each of the papers cited. They show that a great number of subtle changes are brought about in plants by water stress that cumulatively are seen as more obvious gross changes—for example, senescence of leaves, reduced growth, and early maturation. This suggests that it is highly improbable that any single parameter can adequately reflect the general response of the plant to water stress, whether for a short stage of development or for the entire ontogeny of the plant. Even if a number of parameters is available, there are further difficulties for the plant breeder, which can be illustrated by a hypothetical example as follows:

A plant breeder is faced with an environment that produces on average a period of water stress in a species for 2 weeks before anthesis, and this stress demonstrably leads to severe yield reduction. The options for breeding a new plant that could tolerate this period of stress without loss of yield might be as follows:

(a) *The cognitive approach.* To avoid the natural seasonal variability of field conditions, the breeder would first reproduce, on a small

TABLE 22: A Cross Section of Current Papers in the Area of Physiological Response to Drought Stress

Species	Plant Response to Water Stress
Wheat (Triticum aestivum)	Chlorophyll formation, proline, and abscisic acid Osmotic potential, elastic modulus, and apoplastic water Loss of K^+ from roots Chromatin activity Nitrate and nitrite reduction in leaves Seed set Solute accumulation Abscisic acid Polyribosome content Shoot and root growth Fatty acid content
Rice (Oryza sativa)	Inorganic pyrophosphatase activity Leaf water potential and leaf water content Xylem water potential Assimilate translocation Leaf-rolling, stomatal resistance, leaf water potential Respiration
Maize (Zea mays)	Nitrogen metabolism Seed quality Photosynthesis and photorespiration Resistance to CO_2 assimilation Dry-matter production & translocation Exosmosis Proline accumulation Nitrate reductase Carbohydrate metabolism Solute regulation Sulfhydryl dependent reactions Nutrient accumulation Net assimilation and leaf-area index
Barley (Hordeum vulgare)	Photosynthesis Leaf area and photosynthesis Soil CO_2 production Amino acid translocation
Oats (Avena sativa)	Cuticular transpiration rate Leaf diffusion resistance

Reference

Bengston, Klockare, Klockare, Larsson, & Sundqvist. Physiol. Plant. 43: 205 (1978).
Campbell, Papendick, Rabie, & Shayo-Ngowi. Agron. J. 71: 31 (1979).

Erlandsson. Physiol. Planta 47: 1 (1979).
Genkel, Satarova, & Shaposhnikova. Soviet Plant Physiol. 26: 336 (1979).
Heuer, Plant, & Federman. Physiol. Plant. 46: 318 (1979).
Morgan. Nature (Lond.) 285: 655 (1980).
Munns, Brady, & Barlow. Aust. J. Plant Physiol. 6: 379 (1979).
Quarrie. Nature (Lond.) 285: 612 (1980).
Scott, Munns, & Barlow. J. Exp. Botany 30: 905 (1979).
Sepaskhah & Boersma. Agon. J. 71: 746 (1979).
Willemot & Pelletier. Can. J. Plant Sco. 59: 639 (1979).

Dwivedi, Kar, & Mishra. Irrigation Sci. 1: 119 (1979).
Hirasawa & Ishihara. Jap. J. Crop Sci. 47: 655 (1978).
Ishihara & Hirasawa. Plant and Cell Physiol. 19: 1289 (1978).
Kobata & Takami. Jap. J. Crop Sci. 48: 75 (1979).
O'Toole & Cruz. Plant Physiol. 65: 428 (1980).

Volkova & Motkalyuk. Fiziol. Rast. (Mosc) 25: 1244 (1978).

Botha. J. S. Afr. Botany 46: 45 (1980).
El-Forgany & Makus. J. Am. Soc. Hortic. Sci. 104: 102 (1979).
Fock & Lawlor. Ber. Deutch. Bot. Ges. 92: 145 (1979).
Goudriaan & Van Laar. Photosynthetica (Prague) 12: 241 (1978).
Jurgens, Johnson, & Boyer. Agron. J. 70: 678 (1978).
Kostic & Popovic. Biol. Vestn. 25: 178 (1977);
Mukherjee. Plant Cell Physiol. 21: 197 (1980).
Popovic & Kostic. Biol. Vestn. 25: 191 (1977).
Thakur & Rai. Biol. Plant. 22: 50 (1980).
Sharp & Davies. Planta 147: 43 (1979).
Tomati & Galli. J. Exp. Botany 30: 557 (1979).
Verasan & Phillips. Agron. J. 70: 613 (1978).
Wilson & Allison. Rhod. J. Agric. Res. 16: 175 (1978).

Leach. J. Agric. Sci. 94: 623 (1980).
Legg, Day, Lawlor, & Parkinson. J. Agric. Sci. 92: 703 (1979).
Pritchard. J. Sci. Food Agric. 30: 547 (1979).
Tully & Hanson. Plant Physiol. 64: 460 (1979).

Bengtson, Larsson, & Liljenberg. Physiol. Plant. 44: 319 (1978).
Sandhu & Horton. Agric. Meteorol. 19: 329 (1978).

(continued)

TABLE 22: *(Continued)*

Species	Plant Response to Water Stress
Soybean *(Glycine max)*	Leaf expansion Leaf water potential and diffusive resistance Photosynthesis and nitrogen fixation Yield components Leaf water potential–water extraction Xylem water potentials and growth Stomal conductance and leaf water potential Glycolic acid oxidase activity
Sunflower *(Helianthus annuus)*	Resistance to water movement Potassium loss in guard cells Photosynthesis and dry-matter partitioning Osmotic adjustment Photosynthesis and wilting Hill reaction activity
Rapeseed *(Brassica* spp.*)*	Photosynthesis Pigment changes Nitrate metabolism Inhibition of germination Change in gibberellin levels
Beans *(Vicia* and *Phaseolus* spp.*)*	Fertility and seed index Peroxidase isoenzymes Callose formation in anthers Photosynthetic activity Potassium uptake and stomatal aperture Leaf O^{18} enrichment Nitrogen metabolism Photosynthesis and translocation Leaf number and area Enzyme activity Carbon dioxide exchange Rooting depth RNA synthesis Diffusive conductance of epidermis
Peas *(Pisum sativum)*	Root formation Leaflet movement
Peanut *(Archis hypogea)*	Organic acid metabolism Water potential

Reference

Bunce. Can. J. Bot. 56: 1492 (1978).
Carlson, Momen, Arjmand, & Shaw. Agron. J. 71: 321 (1979).
Fellows, Patterson, Gross, & Harris. Plant Physiol. 63: 139 (1979).
Momen, Carlson, Shaw, & Arjmand. Agron. J. 71: 86 (1979).
Reicovsky & Deaton. Agron. J. 71: 45 (1979).
Scott & Geddes. Weed Sci. 27: 285 (1979).
Sivakumar & Shaw. Iowa State J. Res. 54: 17 (1979).
Tashakorie, Thi, and Silva. Compt. Rend. Hebdomadaires Seances Acad. Sci. Sec. D. 289: 1101 (1979).

Black. J. Exp. Botany 30: 235 (1979).
Ehret & Boyer. J. Exp. Botany 30: 225 (1979).
English, McWilliam, Smith, & Davidson. Aust. J. Plant Physiol. 6: 149 (1979).
Jones & Turner. Aust. J. Plant Physiol. 7: 181 (1980).
Rawson. Aust. J. Plant Physiol. 6: 109 (1979).
Volodarskii & Bystrykh. Sel'skokhoz. Biol. 8: 652 (1973).

Baclawska-Krzeminska. Hodowla Rosl., Aklim. Nasien. 17: 303 (1973).
Gross & Lenz. Gartenbauwiss. 44: 159 (1979).
Gupta & Sheoran. Phytochem. 18: 1881 (1979).
Hegarty & Ross. Ann. Botany (Lond.) 42: 1223 (1978).
Rai, Joshi, & Banyal. Experientia (Basel) 34: 1148 (1978).

Abdalla & Fischbeck. Zeit. Acker Pflanzenbau 147: 81 (1978).
Alekseeva & Ramazanova. Dokl. Akad. Nauk SSR 209: 235 (1973).
Barskaya, Balina, & Kanash. Dokl. Akad. Nauk SSR 208–210: 3 (1973).
Chikov & Nikolaev. Fiziol. Rast. (Moscow) 22: 587 (1975).
Cooper & Cockburn. J. Exp. Botany 30: 913 (1979).
Farris & Strain. Radiat. Environ. Biophys. 15: 167 (1978).
Frota & Tucker. J. Soil Sci. Soc. America 42: 743 (1978).
Hoddinott, Ehret, & Gorham. Can. J. Botany 57: 768 (1979).
Karamanos. Ann. Botany (Lond.) 42: 1393 (1978).
O'Toole & Ozbun. Hortscience 9: 281 (1974).
Pospisilova, Ticha, Catsky, & Solarova. Biol. Plant. (Prague) 20: 368 (1978).
Rowse and Barnes. Agric. Meteorol. 20: 381 (1979).
Sheoran & Sihag. Biol. Plant (Prague) 20: 392 (1978).
Solarova & Pospisilova. Biol. Plant. (Prague) 21: 446 (1980).

Rajagopal & Andersen. Physiol. Plant. 48: 144 (1980).
Shackel & Hall. Aust. J. Plant Physiol. 6: 265 (1979).

Singh & Prasad. Geobios (Jodhpur) 7: 14 (1980).
Pallas, Stansell, & Koske. Agron. J. 71: 853 (1979).

(continued)

TABLE 22: *(Continued)*

Species	Plant Response to Water Stress
Lentil *(Lens esculenta)*	Water-use efficiency
Cotton *(Gossypium hirsutum)*	Stomatal response Reduction of auxin transport Adaptive responses Respiration, hexose phosphate accumulation Photosynthesis Acid-soluble phosphorus compounds Stomatal response under nitrogen deficiency Lipid and wax synthesis in leaves
Forage grasses	Root absorption and transpiration Recovery growth Phenology and carbohydrate storage Photosynthesis and transpiration Growth parameters Nitrogen levels Carbohydrate level Rate of leaf extension Energy balance and photosynthetic efficiency Leaf water potential
Forage legumes	Nitrogen fixation, stomatal response Carbon dioxide exchange Stomatal activity Tillering and internode number
Conifers	Changes in growth substance levels Leaf conductance Photosynthesis Evapotranspiration Phenology and photosynthetic rate Stomatal control Abscisic acid and stomatal action Assimilation and root exudation of C^{14} Exudation of carbohydrates by roots
Deciduous trees	Membrane fluidity Osmotic potential, turgor, and growth Water potential Stomatal action Stomatal response to humidity Stomatal resistance, water and osmotic potential Cold hardening

Reference

Yusuf, Singh, & Dastane. Ann. Arid Zone 18: 127 (1979).

Ackerson. Plant Physiol. 65: 455 (1980).
Davenport, Morgan, & Jordan. Plant Physiol. 65: 1023 (1980).
McMichael & Hesketh. Beltwide Cotton Product. Res. Conf. p. 56 (1978).
Nazirov & Tashmatov. Dokl. Vses. Akad. Sel'Skohoz. Nauk. p. 6 (1973).
Parsons, Phene, Baker, Lambert, & McKinion. Physiol. Plant. 47: 185 (1979).
Petinov, Samiev. & Marfina. Fiziol. Biokhim Kul'T. Rast. 6: 9 (1974).
Radin & Parker. Plant Physiol. 64: 499 (1979).
Weete, Leek, & Peterson. Plant Physiol. 62: 675 (1978).

Burch & Johns. Aust. J. Plant Physiol. 5: 859 (1978).
Chu, McPherson, & Halligan. Aust. J. Plant Physiol. 6: 255 (1979).
Cox & Fisser. Arid Land Plant Resources (Texas) p. 529 (1979).
Doley & Trivett. Aust. J. Plant Physiol. 4: 539 (1974).
Francois & Renard. Oecologia Plant. 14: 417 (1979).
Hebblethwaite & McLaren. Grass Forage Sci. 34: 221 (1979).
Horst & Nelson. Agron. J. 71: 559 (1979).
Keatinge, Stewart, & Garrett. J. Agric. Sci. 92: 175 (1979).
Krishnan & Sastri. Arch. Meteorol. Geophys. Bioklimatol. Ser. B. 27: 95 (1979).
Maxwell & Redmann. Physiol. Plant. 44: 383 (1979).

Aparicio-Tejo, Sanchez-Diaz, & Pena. Physiol. Plant. 48: 1 (1980).
Baldocchi & Rosenberg. Bull. Amer. Meteorol. Soc. 59: 1514 (1978).
Ludlow & Ibaraki. Ann. Botany (Lond.) 43: 639 (1979).
Perry & Larson. Crop Sci. 14: 693 (1974).

Bacon & Bachelard. Austral. Forest Res. 9: 241 (1979).
Beadle, Jarvis, & Neilson. Physiol. Plant. 45: 158 (1979).
Brix. Can. J. Forest Res. 9: 160 (1979).
De Velice & Buchanan. N.M. Acad. Sci. Bull. 18: 5 (1978).
Hinckley, Dougherty, Lassoie, Roberts, & Teskey. Amer. Midland Naturalist
102:307 (1979).
Kaufmann. Can. J. Forest Res. 9: 297 (1979).
Newville & Ferrell. Can. J. Botany 58: 1370 (1980).
Reid & Patrick. Plant Physiol. 54: 44 (1974).
Tesche. Flora (Jena) 163: 26 (1974).

Carter, Braden, & Parsons. Hortscience 14: 23 (1979).
Davies & Lakso. Physiol. Plant. 46: 109 (1979).
Jones & Higgs. J. Exp. Botany 30: 965 (1979).
Lakso. J. Amer. Soc. Hort. Sci. 104: 58 (1979).
Schulze & Kueppers. Planta (Berl.) 146: 319 (1979).
Syvertsen & Albrigo. Plant Physiol. (Supp.) 63: 88 (1979).
Yelenosky. J. Amer. Soc. Hort. Sci. 104: 270 (1979).

scale, the general characteristics of the stress-inducing environment (SE) and then determine, using a susceptible genotype, the most obvious physiological and growth and developmental changes that occur during the stress. This would be aided by comparison with a nonstressed environment. Armed with a list of the negative attributes induced in the plant by stress, the breeder must somehow decide which are the alternative plant attributes that will offset the effects of the environmental stress. (An alternative could be to redesign the plant to avoid the stress period close to anthesis by delaying vegetative growth.) The strategy at this point for developing tolerance to the stress would be based on deductions from previously accumulated physiological knowledge for the species in question. The next step would be to locate each of the desirable attributes from the global population and combine them to form a new drought-tolerant genotype. This would require a screening procedure for each attribute and thus the continued use of SE. The steps for self- and cross-pollinated plants could be different at this point, but essentially the procedure would involve a combination of the desirable attributes and a further selection over a number of generations to recover the new plant in either a homozygous form or as a relatively homogeneous population of heterozygous plants. At this point, there would be no guarantee that tolerance to the stress would automatically produce normal yields or maintain quality, so that further selection or crossing might be required to restore these important characteristics of the plant.

(b) *The pragmatic approach.* An alternative and apparently simple strategy would be to simulate the natural 2-week stress and use this artificial sieve to screen that world collection of the particular species for plants that both tolerate the stress and produce normal yields. This strategy is likely to be the first adopted because it does not require a lot of ancillary physiological knowledge. But on the other hand it could fail simply on the grounds that (1) it is physically impossible to screen all the range of genotypes within a species, and (2) the frequency of occurrence of a natural genotype possessing both tolerance to the 2-week water stress and normal yields may be so low as to render it improbable that the selection procedure will recover it.

In any event, strategies (a) and (b) are both very lengthy procedures and depend on the continued use of an artificial-environment screen that simulates accurately the average field condition that induces water stress. The alternative to strategies (a) and (b) could be irrigation to offset the environmental stress. However, where irrigation is not possible, strategy (b) could be invoked where a crop is uneconomical because of its poor stress tolerance and reduced yields. The more difficult cognitive approach would be used last. It is difficult to assess which of the two approaches give a higher probability of ultimate success. In either case, there is a long procedure which will only recover plants that tolerate the specific environmental stress chosen at the beginning of the program. Devising a suitable artificial-environment screen is only possible in the first place if both physiological and growth changes, together with observations of the natural field environment (soil and meteorological) have been closely monitored throughout the life cycle of the plant. Only then does it become clear when water stress occurs and what the nature of the stress is, first in the environment and then within the plant. The advances in such monitoring (Monteith 1975b, 1976) have been considerable in recent years and underpin the approaches that must be adopted if plant breeders are to be successful in breeding for drought resistance.

Techniques

Techniques for simulating water stress in plants vary in their usefulness to plant breeders. Selection of stress-resistant genotypes, whether from a general population of plants or from the F_2 and subsequent generations following crossing, requires the screening of large populations, generally many thousands of plants. Since selection for stress resistance at the later stages of the life cycle is very important, plants need a great deal of space for long periods of time. This tends to rule out controlled-environment chambers and even greenhouses for routine screening because of the cost, particularly in a temperate climate on a year-round basis. The simplest approach, therefore, is to use irrigated plots in a location of very low rainfall on light soils, typically a semiarid or arid climate. A good example is the CYMMYT field station at Sonora, Mexico. Water can be withheld at the particular stage of growth to be investigated. Unpredictable rain showers can be kept off small plot areas by the use of rain-out shelters (Arkin, Thompson, and Chaison 1976; Day et al. 1978; Legg et al. 1978; Teare, Schimmelpfennig, and Waldren 1973) which automatically roll over the plots in response to a sensor system at the onset of rain. A cheaper alternative is to place gutters of asphalt paper between the rows of plants to remove about 50% of the incident rainfall from penetrating the soil, which can be sufficient to create a reasonable level of stress, at least in a semiarid environment (Govinden 1977).

The weaknesses in these techniques are the lack of precise control and the inability to create a stress period at any particular phase of growth by controlling both the soil moisture and the conditions for evapotranspiration. The more elegant the methods of controlling both of these major variables, the smaller the space available and the higher the cost. This works against the needs of the plant breeder. There has been a tendency in physiological experiments to work with small seedlings in pots in order to gain sufficient replication and at the same time conserve space in expensive controlled-environment chambers. The rationale has even been advanced that detecting drought resistance at the early seedling stage will be helpful to plant breeders because this will conserve space and permit looking at more generations instead of waiting the full length of the life cycle. These may be erroneous assumptions since seedling resistance to stress may not necessarily bear any relation to resistance to stress at later stages of growth. The latter under field conditions is usually more significant in reducing yields and quality than the early stages. Under natural conditions, seedlings commonly have optimal soil moisture conditions for growth.

A number of methods are now available for inducing water stress in plants grown under controlled-environment conditions (Slavik 1974). Most methods invoke the restriction of water uptake, either by directly reducing the supply of water to the soil or synthetic root-support medium, or by using an osmoticum such as polyethylene glycol or mannitol to lower the water potential (Lawlor 1969; Slavik 1974). Cooling roots (Kramer 1969) to restrict water uptake can induce mild stress. A novel approach is that of Perroux (1979) where stress to about -11 bar can be accurately induced in a closed system without altering temperature or the composition of the nutrient medium.

The Plant Breeder's Approach

Currently, the number of annual publications by plant physiologists investigating the nature of the responses of plants to water stress is about 3 times the number reporting approaches to improving drought resistance (tolerance and avoidance) in plants of economic interest (Table 23). Most of these latter approaches are being made by what could be described as a new generation of breeder–physiologists. They are concerned primarily with locating genetically determined drought resistance and secondarily with understanding the physiological basis for drought resistance.

Plant breeders seem convinced there is such a genetic trait as resistance to water stress (if judged only by the number of times drought resistance or tolerance is used as a key word in titles of publications). What they are looking for could be put simply as survival, coupled with significant yielding ability, in the face of moderate to severe drought under field conditions. A close study of the breeder approaches (Table 23) suggests that there is need for breeders and physiologists to team up in approaching the difficult practical problem of breeding for drought resistance.

Water Stress and Disease

A number of reports indicate important associations of water stress in plants with the susceptibility to various pathological diseases (Chinnici and Peterson 1979; Colbaugh and Endo 1974; Melchior and Morehart 1979). On the other hand, wilting induced by disease can cause severe damage under conditions of environmental water stress (Jones 1928). Certain physiological disorders arising from prolonged water stress give symptoms similar to disease (Van der Boon 1973) or predispose the plant to susceptibility to other environmental stresses (Markowski et al. 1974).

IN CROP PRODUCTION

Irrigation

Irrigation is the most practical means yet devised for mitigating or preventing drought stress in plants. As early as 5000 B.C., water from the Nile was used for irrigation. China, Peru, and Mexico confirm the antiquity of this practice, and the very complex irrigation schemes of California, Israel, and India provide modern examples of how pastures, field and fiber crops, vegetables, and orchard crops can be grown in arid or semiarid environments severely limited by rainfall and with high potential evapotranspiration rates. These environments, given water, are otherwise suited to plant growth on a year-round basis. It is estimated (Obeng 1975) that of the world's 3.2 billion hectares of potentially arable land, 2,020 million hectares are located in the developing regions of Africa, Asia, and Central and South America; and, of these, 1,330 million hectares constitute irrigable land. It can therefore be anticipated that irrigation, in a variety of forms, will increase on a global scale. For example, in

TABLE 23: Research Directions of Breeder–Physiologists Attempting to Incorporate Some Aspect of Drought Resistance into Crop Plants

Species	Approach to Drought Stress	Indexes
Wheat (*Triticum* spp.)	Resistance	—
	Resistance	Leaf water and osmotic potential
	Resistance	Morpho–physiological traits
	Resistance	Stomatal behavior
	Resistance	Regression analysis of genotype–environment interactions.
	Resistance & sensitivity	Nutrient uptake
	Resistance	Abscisic acid content
	Resistance	Stomatal size and frequency
	Resistance	Stomatal aperture, photosynthetic rates and root pattern
	Tolerance	Seedling emergence
Rice (*Oryza sativa*)	Avoidance	Escape mechanisms
	Resistance	Chlorophyll stability
	Resistance	Mutagenized populations
	Tolerance	Review of available indexes
	Resistance & tolerance	Leaf water potential, leaf rolling, tip drying
	Resistance	—
Pinus spp.	Resistance	Water potential
	Resistance	Natural selection and needle succulence
Tobacco (*Nicotiana tabacum*)	Resistance	Seed germination in mannitol
Tomato (*Lycopersicon esculentum*)	Tolerance	—

Reference

Adjei & Kirkham. Euphytica 29: 155 (1980).

Fischer & Sanchez. Aust. J. Agric. Res. 30: 801 (1979).

Fischer & Wood. Aust. J. Agric. Res. 30: 1001 (1979).

Jones. *In* Mussell & Staples "Stress physiology in Crop Plants," pp. 407–428. Wiley. New York. 1979.
Keim & Kronstad. Crop Sci. 19: 574 (1979).

Kirkham. Phyton. 38: 137 (1980).
Quarrie. Nature (Lond.) 285: 612 (1980).
Tanzarella & Bianco. Genet. Agrar. 33: 355 (1980).

Townley-Smith & Hurd. *In* Mussell & Staples "Stress Physiology in Crop Plants," pp. 447–464. Wiley. New York. 1979.

Zagdanska & Pacanowska. Biol. Plant. (Prague) 21: 452 (1980).

Alluri, Vodouhe, Treharne, & Buddenhagen. *In* Buddenhagen "Rice in Africa," pp. 275–278. Academic Press. London. 1978.
Bhiravamurty & Prasad. Geobios (Jodhpur) 6: 144 (1979).
Gangadharan & Misra. J. Nucl. Agric. Biol. 8: 30 (1979).
Lal & Moomau. *In* Buddenhagen "Rice in Africa," pp. 285–292. Academic Press. London. 1978.
O'Toole & Moya. Crop Sci. 18: 873 (1978).

Reyniers & Jacquot. Agron. Tropicale 33: 314 (1978).

Dykstra. B.C. For. Serv. Res. Notes 62:1 (1974).
Kandyka, Venator, & Mullerstael. Ann. Arid Zone. 18:1 (1979).

Lakshminarayana, Patel, & Jaisani. Indian J. Agric. Sciences 49: 818 (1979).

Richards. Hortscience 14: 121 (1979).

(continued)

TABLE 23: *(Continued)*

Species	Approach to Drought Stress	Indexes
Potato *(Solanum tuberosum)*	Tolerance	Morphological characters
Manihot spp.	Tolerance	—
Grapevine *(Vitis vinifera)*	Resistance	Abscisic acid, stomatal characteristics, water content
Grasses	Resistance Tolerance	Morphological characters Epicuticular wax ultra-structure
Rye *(Secale cereale)*	Resistance	Proline test
	Tolerance	Chromosome identity
Miscellaneous	Resistance Resistance	Mutagenesis —
Maize *(Zea mays)*	Resistance	Chromosome length
	Tolerance	Proline accumulation and potassium response
	Resistance	Biochemical test and field trials
	Resistance	Proline test
	Resistance	Respiration and water status
Barley *(Hordeum vulgare)*	Resistance	Proline accumulation
Soybean *(Glycine max)*	Resistance	Root and shoot water potential
	Resistance	Leaf lamina expansion
	Resistance	Stomatal closure and root growth

Reference

Steckel & Gray. J. Agric. Sci. 92: 375 (1979).

Nassar. Can. J. Plant Sci. 59: 553 (1979).

Fregoni, Scienza, & Miravelle. Grapevine Genetics and Breeding. Second Int. Sympos. Grapevine Breeding. Bordeaux, France. June 14–17, 1977. pp. 287–296 (1978).

Dernoeden & Butler. Hortscience 14: 511 (1979).
Hull, Wright, & Bleckmann. Crop Sci. 18: 699 (1978).

Palfi, Nemeth, Pinter, & Kadar. Acta Univ. Szeged Acta Biol. 24: 39 (1978).

Waines, Ting, & Lazzaro. Genetics 91: S134 (1979).

Mugnozza & Monti. Genet. Agrar. 33: 331 (1979).
Genkel. S-Kh. Biol. 14: 316 (1979).

Chokairi & Gorenflot. Rev. Cytol. Biol. Veg. Bot. 2: 365 (1979).

Mukherjee. Plant Cell Physiol. 21: 197 (1980).

Pinter & Kalman. Maydica 23: 121 (1978).

Pinter & Kalman. Maydica 24: 155 (1979).
Tregubenko & Filippov. Vishnevskii Fiziol. Biokhim. Kul't. Rest. 10: 257 (1978).

Hanson & Nelsen. Crop Sci. 19: 489 (1979).

Sammons & Peters. Crop Sci. 18: 1050 (1978).

Sammons, Peters & Hymowitz. Crop Sci. 19: 719 (1979).
Wien, Lyttleton, & Ayanaba. In "Stress Physiology in Crop Plants," pp. 283–302. Wiley. New York. (1979).

(continued)

TABLE 23: *(Continued)*

Species	Approach to Drought Stress	Indexes
Cotton *(Gossypium hirsutum)*	Resistance	Proline accumulation and growth
Winged beans *(Psophocarpus tetragonolobus)*	Tolerance	Morphological characters
Tepary beans *(Phaseolus acutifolius)*	Tolerance	Protein content and grain weight
Cowpea *(Vigna* spp.*)*	Resistance	Stomatal closure, photosynthesis
Peas *Pisum sativum)*	Tolerance Tolerance	Osmotic properties Protoplasmic viscosity

recent years there have been major increases in India and Bangladesh with the adoption of tube-wells (Appu 1974); the major surface-water development already has an installed capacity for irrigating 20 million hectares (potential of 65 million hectares) in the total geographical area of 328 million hectares (Barton Worthington 1976).

Successful use of irrigation depends on basic knowledge in three areas. (1) When and how much to apply water to supplement the natural evaporation and evapotranspiration. (2) A knowledge of the water-absorbing and -holding capacity of the various soils together with ways of preventing and overcoming physical problems such as water erosion, salinity, and nutrient deficiency. (3) Knowledge of plant functions and growth in the presence of soil water in optimum and less than optimum amounts. The many hundreds of papers published in the last 80 or so years about a wide range of plant species in a great range of environments, which compare the growth of irrigated and rainfed crops, bear witness to the search for this basic information.

Early research into the growth and water-relations responses of plants was concerned with water-use efficiency and major characteristics such as yield, because of their direct relationships to the

Reference

Ferreira, De Souza, & Prisco. Z. Pflanzenphysiol. 93: 189 (1979).

Karikari. Acta Hortic. (The Hague) 84: 19 (1977).

Waines. Crop Sci. 18: 587 (1978).

Wien, Lyttleton, & Ayanaba. *In* "Stress Physiology in Crop Plants," pp. 283–302. Wiley. New York. (1979).

Stadelmann & Stadelmann. Plant Physiol. 63: 87 (1979).
Stadelmann, Young, & Stadelmann. Hortsci. 14: 27 (1979).

economic costs of applying water (Russell 1959). In recent years, emphasis has shifted to plant architecture as a means of optimizing the capture of light by leaves and the recovery of nutrients and water by roots (Borlaug 1968; Clarke and Simpson 1978). When water deficits are removed, light and nutrients become major limiting factors to plant growth.

At first sight, the question of when and how much to irrigate seems a simple problem of (a) knowing how much water is present in the soil, (b) knowing the rate of evaporation from the soil, and (c) replacing the water at specific intervals, or continuously, to maintain an optimum set of water conditions for plant growth without serious or irreversible stress effects in interim periods. Both (a) and (b) can be determined by relatively simple meteorological observations which involve determination of rainfall and evaporation, radiation fluxes, and windspeed. The theoretical background to (c) is well established (Penman 1949; Thornthwaite and Holzman 1939) and put to some considerable practical use, but it is nevertheless fraught with difficulties. The answer to the question of what is the optimum soil water status for plant growth varies with the nature of the soil water-holding capacity and degree of availability to the plant, properties that differ considerably between light and heavy soils.

The main factors controlling the availability of water to a crop are the soil water diffusivity (the ratio of the hydraulic conductivity to the specific water capacity of the soil; the root density–depth distribution; root diameter; plant resistances to water movement; the leaf water potential at which pronounced stomatal closure occurs; and the potential evaporation from the crop) (Sedgley et al. 1973). It is also difficult to predict, from meteorological observations, the actual transpiration of the plant (compared with actual evaporation from the uncovered soil surface) because of the ability of plants to reduce evapotranspiration by closing stomata, by reducing leaf area through senescence, by development of waxy cuticles, and through leaf-rolling, increased leaf osmotic potential, and reduced root growth. In effect, evaporation and transpiration can be regarded as two independent variables (Denmead 1973).

Stanhill (1976) summarized this current position of the great need for knowledge of the internal control of plant water relations by reference to an important irrigated crop, cotton. He showed that cotton is similar to other mesophytic crop species with respect to its transfer properties in relation to the environment; that is, the transfer of radiation, momentum, and heat can be estimated with considerable accuracy from climatological data but not without the addition of a single term describing the size of the stand. Stanhill went on to note that, given additional information on the internal resistance to diffusion, it is possible to measure water-vapor transfer with sufficient accuracy for field applications. However, he emphasized that while the complete meteorologically based model is of great agronomic interest, the limiting factor at present is the basic lack of understanding of the effect of the climate on the size of the stand and its internal resistances, further compounded by the lack of information on the functional relationships between the development and relative growth of the reproductive and vegetative tissue and the climate.

Perhaps the most difficult area is understanding the role of root growth under drought stress, because of the technical difficulties of observing roots in situ. Many ingenious devices have been constructed to observe root growth under controlled-environment or greenhouse conditions (Iowa State University 1980; Taylor and Klepper 1978). Extrapolating this knowledge to field-grown plants is difficult. Root growth in large bodies of undisturbed soil, in plants exposed to the more rigorous outdoor environmental fluxes, is considerably different from that under artificial environments. The onset of seasonal stress under field conditions—particularly on plants grown in heavy soils—is slow, and plants are able to adjust. One of the practical difficulties in controlled-environment experiments which aim specifically at controlling soil moisture tension—a major determinant of drought stress (Weatherley 1965a)—is a way of altering consistently the soil/hydroponic medium water potential. The literature is full of reports where water is "withheld to cause water stress," often in small plots. There is no doubt that plants can grow normally in well-watered pots of soil, but this type of root environment is inadequate for studying water stress, if only because of the uncharacteristic way small bodies of soil dry out (the outside layers of soil where most of the roots congregate dry abnormally quickly).

Probably the most useful technological developments that have depended on the advancement of physiological knowledge about drought stress are trickle irrigation (Benasher 1979; Freeman et al. 1976; Singh and Singh 1978) and the use of saline water (Jury et al. 1978; Mondal and Sharma 1979; Singh et al. 1978) for trickle irrigation, two techniques that have opened up new possibilities for certain types of crop production in the arid zones.

Rainfed Conditions of Agriculture

Two aspects of improving plant growth under rainfed conditions, particularly of the semiarid and arid zones, are the possibilities of deliberately extending the range of existing species of plants in terms of their adaptability to drought stress and also introducing new species with improved drought resistance for particular zones. The extension of desert landscape to areas that once did not have these characteristics (desertification) has increased dramatically in the last 25 years due directly to a demographic explosion (Secretariat U.N. 1977) (see next section for further discussion). In North Africa, desertification is occurring at an accelerated pace (for example, 820,000 ha yr^{-1} in the Sudan), and satellite pictures indicate increased deserts in the Sahel, Iraq, Jordan, Saudi Arabia, and Syria. Estimates indicate that one-tenth of South America, one-third of Asia and Africa, and one-quarter of Australia are in danger of desertification. Two of the five major factors contributing to the extension of deserts are the extension of crop growing into ecologically unsuitable zones and the salinization and alkalinization of soils due directly to poor irrigation. Paradoxically, then, irrigation used with mesophytic plants to maximize crop yield in arid environments can have serious negative effects on the long-term use of the environment. A strong case can therefore be made for selective improvement of drought adaptation in plants that are already present in the arid and semiarid zones (Felger and Nabham 1976). Grasses are generally dominant species in arid and semiarid zones; and, while there is a steady output of physiological research on grasses, most of it is on the temperate zone grasses, typically with mesophytic characteristics. There is thus a need for a significant increase in physiological–genetic studies of grass and legume species that currently attract little interest because of their apparently low economic value but which in future may provide the only hope for stabilizing the sensitive, drought prone, arid and semiarid zones.

Stomata play a major role in reducing excessive transpiration during diurnal and intermittent daily periods of drought stress (Raschke 1975a). Since the discovery that naturally occurring plant-growth regulators such as abscisic acid (Mansfield et al. 1978) can control stomatal action, there has been a large increase in research into the general mode of action of natural and synthetic antitranspirants with a view to their practical use in reducing evapotranspiration in both irrigated and nonirrigated plants (Das and Raghavendra 1979). While many of the synthetic compounds such as phenyl mercuric acetate (Miller and Ashby 1978) and substituted alpha-hydroxy sulfonates (Gacso and Lassanyi 1978) are very effective antitranspirants by their ability to shut stomata, the practical difficulties associated with determining exactly when and how to apply the compounds, and how long to reduce stomatal transpiration, have prevented significant practical use. In addition, many of the most effective compounds are hazardous to the environment or to humanity. Nevertheless, a great deal of valuable information has been obtained about the role of stomates in drought resistance by the use of these compounds.

Another approach to altering drought resistance has been investigation of the possibility of altering plant morphology by the use of growth regulators. There has been little success, despite widespread investigations for nearly 30 years, including into the use of such potent growth inhibitors as tri-iodobenzoic acid (TIBA) and 2-chloroethyl trimethyl ammonium chloride (CCC) (Chrominski 1972; Henckel and Pustovoitova 1977), which have been the subject of much investigation in eastern Europe and the Soviet Union. These two particular compounds are currently used on quite a large scale in western Europe to prevent lodging, by their dwarfing action, of mesophytic temperate plants (Heyland et al. 1975; Jung and Dressel 1973).

Another area of significant interest has been that of "pre-sowing" and other forms of hardening against subsequent exposure to drought stress. Considerable claims have been made by Soviet workers (Filatov and Frolova 1972; Genkel et al. 1979; Tarabrin and Teteneva 1979). How much of the observed modifications of drought resistance (Filatov and Frolova 1974; Srivastava and Singh 1979) can be attributed to changes in plasmatic hardiness, cellular tolerance to water stress and dehydration, stress avoidance due to induced anatomical changes such as leaf senescence, reduced leaf or tiller formation, and so forth is unclear in much of the published work, particularly the Soviet work which often gives scanty detail of methods and results. An obvious limitation to the use of pre-hardening under field conditions would be that control has to be exerted over moisture availability to plants, which would confine its use to irrigated crops—unless, of course, ways are discovered of inducing the same response by the application of growth regulators.

Probably the greatest single limiting factor to advancement of the application of physiological knowledge to field crop production under drought-stress conditions has been the lack of simple techniques that can be used directly in the field for monitoring the water status of large numbers of plants. Slavik's review (Slavik 1974) summarizes the trends in methodology, primarily for use in the laboratory.

Despite the considerable accuracy that can now be achieved, the limitations for many methods are either that they require removal of plant parts to the laboratory or that the method is time consuming on apparatus that can only deal with a few plants. Some of the methods are only suited for laboratory conditions or plants grown in controlled environments. A cross section of recent papers (Table 24) indicates the diversity of approaches and refinements of older methods in the search for better ways of measuring how the plant is avoiding, tolerating, or succumbing to external water stress.

SUSTAINING A STABLE ECOSYSTEM

A very close correlation exists between vegetation and climate (Walter 1973). While it has seemed fairly apparent how shifts in climate can alter vegetation, it is less obvious how changes in vegetation

TABLE 24: Some Recent Investigations for Physical Techniques for Measuring Plant Water Status

Physical Technique	Characteristic Measured in the Plant
Gasometric system	Evaporation during transpiration
Electronic	Transpiration
Osmometer	Leaf osmotic potential
Pressure volume	Turgor potential and cell elasticity
Spectral analysis	Turgor responses
Electrical conductivity	Desiccation injury
Hydrolysis of statolithic starch	Drought and heat resistance
Two sensitive strain guages	Nondestructive measure of pressure potential
Pressure chamber and diffusive resistance	Leaf water potential
Field lysimeter	Crop evapotranspiration
Monitoring tritiated water	Transpiration rate
Pressure chamber	Leaf water potential
Thermocouple psychrometer	Leaf and xylem water potentials
Dielectric constant at ultra-high frequencies	Water in chloroplasts
Visual estimate	Seedling emergence under drought stress
Visual estimate	Plant water status
Diffusion porometer	Stomatal humidity
Hygrometer	Leaf and stem water potential
Pressure chamber	Water potential of small leaves
Leaf extensiometer	Leaf extension in grasses
Diffusion porometer	Stomatal conductance
Micrometer	Stem diameter changes to measure canopy water potential
Tensiometer–potometer	Water fluxes and potential
Root and soil psychrometer	Water potential gradients

may affect first microclimate and in turn influence macroclimate. Technology has been described as "the difference between ecologic and economic cycles" (Galtung 1979). The extent to which technology has expanded in the last century has created a new dimension for shifts in both climate and vegetation. Vast depredations of forest in West Africa and Brazil for commercial timber and establishment of agriculture; plowing of semiarid landscapes (western North America and Kazakhstan in the Soviet Union); massive overgrazing of arid

Reference

Bittman & Steppler. Can. J. Plant Sci. 59: 545 (1979).
Block & Sterzelmeier. Gartenbauwiss. 43: 142 (1978).
Clarke & Simpson. Agric. Water Manage. 1: 351 (1977).
Cutler, Shahan, & Steponkus. Crop Sci. 19: 681 (1979).
Davis & Huck. Crop Sci. 18: 605 (1978).
Dlugokecka & Kacperska-Palacz. Biol. Plant (Prague) 20: 262 (1978).
Genkel & Shelamova. Sov. Plant Physiol. 25 (Part 2): 151 (1978).

Heathcote, Etherington, & Woodward. J. Exp. Botany 30: 811 (1979).

Hummel, Pellett, & Parsons. Can. J. Plant Sci. 59: 847 (1979).

Hutson, Green, & Meyer. Water Sa. 6: 41 (1980).
Ibrahim, Berger, & Rapp. Plant and Soil 52: 291 (1979).
Ike, Thurtell, & Stevenson. Can. J. Botany 56: 1638 (1978).
Ishihara & Hirasawa. Plant Cell Physiol. 19: 1289 (1978).
Ishmukhametova & Rybkina. Sov. Plant Physiol. 19: 318 (1972).

Johnson & Asay. Crop Sci. 18: 520 (1978).
Jones. J. Agric. Sci. 92: 83 (1979).
Kenny & McGruddy. Agric. Meteorol. 10: 393 (1972).
Pallas & Michel. Peanut Sci. 5: 65 (1978).
Roberts & Fourt. Plant Soil 48: 545 (1977).
Sharp, Osonubi, Wood, & Davies. Ann. Botany (Lond.) 44: 35 (1979).
Sivakumar & Shaw. Agron. J. 70: 619 (1978).
So, Reicosky, & Taylor. Agron. J. 71: 707 (1979).

So, Aylmore, & Quirk. Plant Soil 49: 461 (1978).
So. *In* Harley "Soil–Root Interface," pp. 99–113. Academic Press.
London. 1979.

lands (southern Sahel of Africa); global strip-mining for minerals and coal; and giant irrigation projects and the use of tube-wells and overhead-sprinkler irrigation (Egypt, northern India, and southwestern United States) are some of the kinds of changes that have taken place in short periods on a scale unimaginable before the advent of the industrial age. On top of this, the congregation of the mass of humanity into concentrated urban environments has tended to disperse to and at the same time neglect environmental problems at the geographical periphery distant to urban centers (Waddell 1977). These geographical peripheries tend to be the more fragile arid and semiarid zones of the world, including the Arctic, where plant life is particularly sensitive to small shifts in water balance.

Desertification is a natural process that can be exacerbated, by the activities of man, in a number of ways. Most of these ways relate to increased population density (Secretariat U.N. 1977). Examples are the penetration of agriculture into arid and semiarid zones, denudation of vegetation for fuel (particularly Africa, northern India, and the Middle East), local over-grazing associated with the use of tube-wells (southern Sahel of North Africa), the ruination of soils through faulty irrigation with subsequent wind and water erosion once plant cover is gone, and the too-frequent slash and burn agriculture of the semiarid tropics (North Africa and North Thailand).

There is general debate, and the answer is not clear, as to what extent increased desertification (Secretariat U.N. 1977) and the increased global concentration of CO_2 in the atmosphere (Myers 1970) may contribute significantly to climatic fluctuations. Historically, the concentration of CO_2 in the atmosphere has changed from 290 to 330 ppm, with predictions, based on fossil fuel consumption, of 400 ppm by the year 2000 and 600 ppm by 2050 (Myers 1970), and an accompanying theoretical global increase of temperature of $1°C$. The temperature rise would give a longer growing season and more rainfall in the subtropical arid regions.

Regardless of the degree of causality involved, it is clear that in the long term those vegetation zones already radically altered must be stabilized and hopefully reconstituted. Prevention, rather than cure, would seem to be the wisest and least costly approach, if the experience of Israel in reclothing much of the denuded landscape at enormous cost, is an example to judge from. Many arid-zone areas have been lost permanently to desertification and the once "fertile crescent" of the ancient Middle East is mute testimony to the mismanagement of past civilizations.

The subdiscipline of physiological ecology has grown rapidly as plant physiologists and meteorologists have begun to analyze the interactions between plants and the atmosphere at the level of vegetation or climatic zones. The International Biological Program (1974) was the first serious global attempt to study climate and vegetation on a scale large enough, and with an interdisciplinary approach, that can lead to a better understanding of the global ecosystem. The information yielded from physiological studies of normal plants growing in the arid and semiarid zones is essential for the development of strategies for the reclamation of these zones when they have been radically altered. For example, studies of the natural grassland ecosystem (Ripley and Redmann 1976) and tundra (Lewis and Callaghan 1976) are essential as base-line information for reclamation plans.

One of the significant areas of physiological study that is leading to a clearer understanding of the global ecosystem is "ground-truthing," or base-line studies for interpreting satellite remote sensing of broad vegetation zones (Myers 1970; Wittwer 1978). Remote sensing has already shown the magnitude and rate of desertification (Secretariat U.N. 1977) and can also provide information needed for controlling irrigation in agriculture. Remote sensing is dependent on basic information about the structure and function of the vegetation canopy, which can be correlated with the absorbed or reflected radiation data stored as information incorporated into the computer data bank associated with each satellite. A proper interpretation of remotely sensed data depends on a fundamental knowledge of the basis of reflectance of solar radiation from crop canopies and how reflectance changes with stress conditions of plants. As a vegetative canopy is stressed, growth and development slow, leaves may wilt, changes occur in internal water content, and leaf temperature may increase. All of these effects will alter the reflected spectrum of solar radiation. In the near infrared there is a high reflectivity from plant surfaces (Bartholic et al. 1972; Tucker 1979) which is markedly altered by changes in plant water status accompanying stress. Black and white and color photography can be used for mapping vegetation, and interpretation (Avery 1968) or, alternatively, information can be selectively obtained through various forms of computer printout. Thermal scanning is a new technique (Millard et al. 1978). At present the complexities of interpretation are great but the approach is likely in the long term to be of great value for understanding the global ecosystem.

6

G. M. SIMPSON

The Research Challenges

RESEARCH METHODOLOGY

As with most other subbranches of biological research, the area of drought-stress physiology depends very much on technical advances in instrumentation for monitoring physical and chemical changes in the status of the plant in relation to changes in the environment. Micrometeorological methodology has advanced rapidly in recent years. Multidisciplinary programs such as the International Biological Program (1974), in which international studies on a global basis were carried out on a number of different ecosystems, led to the development of extensive systems of instruments. These systems have been used for monitoring all aspects of the aerial and soil environment in relation to natural associations of plants.

It is quite difficult to monitor a single plant under field conditions. However, a partial way around this problem is the development of sophisticated controlled-environment chambers in which light intensity, spectral quality of light, humidity, day length, temperature, and gaseous composition can be accurately controlled. Plants can be grown throughout their life cycle in a simulated environment under repeatable conditions (Evans 1963; Langhans 1978). In

/

addition, if the plants are genetically identical, the average behavior of a single plant can be observed in a situation where all aspects of the external environment can be defined in relation to specific facets of growth—for example, root or leaf growth.

Direct measurement of functional or structural changes within a plant are difficult without destruction of the integrity of the plant. Most instruments used for monitoring a single intact plant modify the microenvironment of the plant, particularly if the measurements are prolonged. Nevertheless there is now a considerable range of instruments available for making rapid measurements of water status, water content, and water exchange between soil and roots, as well as leaf gas conductance and water conductance within the plant (Slavik 1974). Most of the methods are nondestructive, so that repeated measurements on the same plant are possible. Improvements can be expected in these kinds of instruments through the application of lasers, transducers, ultrasound, nuclear magnetic resonance, and the computer, to name but a few of the approaches currently being explored.

The most difficult area of investigation is the examination of tissues, single cells, and subcellular components in relation to changes in water status. All methods at this level tend to involve considerable alteration in the system, to the extent that it becomes difficult to relate the measured changes to the observed behavior of intact plants. For example, there is currently much interest in the structure and role of plant cell membranes in relation to drought and cold resistance (Lyons et al. 1980). Examination of membranes requires complicated and refined techniques generally involving fragmentation of the membranes into constituent parts. Cellular behavior is interpreted from a synthesis of the behavior of the constituent parts. This may not represent a real behavior of an intact membrane in an intact cell.

Whatever the approach, trying to understand the nature of drought resistance in plants is a problem in systems analysis (Laszlo 1972; Von Bertalanffy 1968) in which the water relations of the plant is only one of the subsystems in the dynamic, multilevelled hierarchical functional system that comprises the plant and its environment. Just as the function of the total system is generally limited, or controlled, by the activity of one of the subsystems, so is our understanding of the total response limited by a lack of information about the subsystem, or subsystems, in some particular level of the hierarchy of functional levels. Thus our view of the nature of drought resistance in plants must include the many other transactions that take place in plants.

Levitt (1978) has pointed out that crop plants growing under any conditions are subjected to a range of sub- and supraoptimal conditions in terms of the range of environmental variables that exist at any one time. A temperature of 25°C may be suboptimal for one aspect of growth but supraoptimal for some other aspect at the same moment in time or at some specific stage of growth of the whole plant. There is thus an important distinction between both sub- and supraoptimal conditions that merely attenuate a function, as against sub- and supraoptimal conditions that cause irreversible and harmful changes which lead to death of parts or ultimately of all the functional system. Levitt has described these two kinds of stress as moderate and reversible and severe and irreversible. It is these two ranges of stress, and plant responses to these stresses, that have become the focus of interest of physiologists interested in drought resistance of plants. Historically, interest has been mostly with the moderate stress effects because of their obvious relationship to crop yield. Irrigation is a practical way of alleviating moderate stresses. However, as irrigation becomes more costly and difficult, attention will have to be focused on the severe stresses that determine nonsurvival and total loss of yield.

Given that the field of drought stress physiology is a fairly active one, it is significant that other fields of interest in stress related to other environmental factors are also very active. Recent publications—for example, in the areas of low- and high-temperature stresses (Levitt 1980a; Li and Sakai 1978; Lyons et al. 1980) and radiation, salt, and other stresses (Levitt 1980b)—suggest the possibility, in the near future, of integrating a number of subfields of stress physiology into a more comprehensive understanding of stress; that is, understanding of the general response of crop plants to stress from the environment.

INTEGRATING THE KNOWLEDGE OF DROUGHT RESISTANCE IN PLANTS WITH CROP PHYSIOLOGY

Our knowledge about all aspects of plants has expanded exponentially since the 1940s as a result of the great increase in the number of trained plant scientists. There is now a great range of knowledge extending from the more holistic level of understanding about the behavior of populations down to the extremely detailed knowledge of subcellular structure and function—for example, of the DNA molecule, enzymes, or hormones. The scale of the volume of this knowledge is staggering. The information produced in the botanical field in 1 year is now more than an individual scientist could read in a lifetime. Computers must be used to search the vast store of information related to any one topic, either to find the latest information or to undertake a retroactive search related to a specific topic. The problem of integrating this knowledge has become one of the central problems of biology.

At the same time, the research approach used most frequently in science has been, until recently, the reductionist approach (Koestler 1967) which creates more knowledge. That is, the subject of interest is systematically dissected into constituent parts which are then further subdivided. The relationships within each of the successive subdivisions are examined. In this way a hierarchy of levels of understanding is created which may depend on structural relationships between levels or functional relationships, for example, with a set of physiological relationships. Looking upward in the hierarchy gives a holistic, integrative view; looking downward gives a more detailed and fragmented view.

Physiologists interested in control mechanisms have been the first to run into the problem of developing useful concepts that integrate the various constituent levels of understanding into a hierarchical systems model that accurately portrays the functional system comprising a plant in its environment. The general concept of homeostasis (Yamamoto 1965), whereby a living organism sustains an open system in which continuous exchange with the environment occurs, is probably the most basic idea useful to systems thinking. Changes in the surroundings excite reactions in the living system or affect it directly so that internal disturbances are produced. Such disturbances are usually kept within narrow limits because automatic adjustments within the system are brought into action, and thus wide oscillations are prevented and internal conditions are held fairly constant. Feedback plays an important part in controlling perturbations in the system.

Higher animals have the ability to regulate their internal environment by a range of mechanisms that permit a high degree of independence of the system from the immediate environment. On the other hand, plants are particularly exposed to a wide range of environmental fluctuations. They have evolved a remarkable range of mechanisms that can confer viability for some particular species throughout a wide scale of environmental conditions. For example, some lichens can remain viable over the temperature range −196 to +50°C. The desiccation tolerance of some lichens can be as low as 1.1% water content for 34 weeks without loss of viability (Bewley 1979).

Response to drought stress in the environment is therefore just one of the many daily or seasonal perturbations of the environment to which plants must constantly adjust. Interpretation of the responses to drought stress thus depends on an understanding of the other variables in the environment. Handling and integrating all the information that relates to the different environmental variables, their interactions with each other, and their interactions with the hierarchy of functions within a plant becomes an exercise in systems analysis. A static model of the plant system is the first approximation to developing a dynamic model where functions can be understood in terms of energy or substrate transactions.

Modeling is essentially putting back together the parts of the system that were separated, to find out how the subparts function. The key to solving the dynamic aspects of biological models is undoubtedly the digital computer (Duncan 1967). The many thousands of transactions and calculations needed to interrelate the different levels of a dynamic model can only be handled by a computer. Modeling quickly shows up the missing gaps in our knowledge and provides the quantitative hypotheses which can be tested in the field or laboratory. The sensible development of a model provides a sound framework for further experimental work, rather than the sporadic serendipitous approach that is the basis of much research.

Related to the central conclusion that understanding the response of plants to drought stress requires a systems approach could be a further conclusion that research in this area should be undertaken by coordinated teams of scientists from a number of disciplines using an interdisciplinary approach. In this way the "pieces" of knowledge can be fitted together to complete a holistic, integrated picture. Coordinated research ensures that the experiments and information acquired are compatible and specifically designed for synthesis into a comprehensive model. The days of scientists working in splendid isolation may well be over. Such a team approach to a drought-stress problem in plants could well involve experts in the areas of soils, meteorology and micrometeorology, agronomy, chemistry, biophysics, physiology, genetics, cytology, biometry, computer science, and engineering, to name some of the most obvious. The application of the information acquired by a systems approach would require further interfacing between the disciplines of agricultural engineering, plant breeding, agricultural extension and communication, sociology and agricultural economics, rural planning and transport, and, most important of all, farmers. The problems of communication brought about by specialized vocabularies used within each discipline is often a major obstacle to cooperative interaction.

APPLICATION OF DROUGHT-STRESS PHYSIOLOGY TO AGRICULTURE

The task of researching and applying information about the response of plants to drought stress is probably analagous to a complicated task such as putting a man on the moon. Once the objective has been defined—for example, improving drought resistance of a specific crop in the semiarid zone of North Africa—a logistical study can map out the subtasks that must be fulfilled. The subtasks will define the kinds of people that will be needed to achieve success in each subpart of the overall problem. Complicated tasks such as building sophisticated aircraft, space exploration systems, and the coordination of government bureaucracies are routinely performed today because of vast improvements in communication systems. The task of increasing the productivity of a single species of crop plant, grown under drought-prone conditions, is an exercise in systems analysis and coordination along similar lines.

Clearly, considerable capital investment is needed to deal with all the tasks, but, without motivation, success is unlikely. The key to success of a multi-tiered problem-oriented research and development project lies largely with the motivation for success of the people involved. The global pressure of increasing population may ultimately provide this motivation, as hunger spreads from the least-developed countries to the currently affluent countries.

There is already a great deal of accumulated knowledge about the response of plants to drought stress. However, most of this knowledge comes from a restricted number of crop plants. There is room for more study among the great range of noncrop plants adapted to the semiarid and arid zones of the world. This is necessary simply to enlarge our understanding of the range of mechanisms of drought adaptation and resistance. There is no doubt that knowledge about water relations in plants is a field of primary significance to agriculture and global food production. The prospects for application of existing knowledge, either directly via agronomy or through plant breeding, are good. The potential returns in terms of stabilized and increased crop production can be very high, as the ancient practice of irrigation testifies. It remains for us to ensure the coordination of sensible approaches and to develop the will to carry out the tasks.

REFERENCES

Abeles, F. B., and Leather, G. R. 1971. Abscission: Control of cellulase secretion by ethylene. *Planta* 97: 87-91.

Acevado, E., Fereres, E., Hsiao, T. C., and Henderson, D. W. 1979. Diurnal growth trends, water potential and osmotic adjustment of maize and sorghum leaves in the field. *Plant Physiol.* 64: 476-80.

Ackerson, R. C. 1980. Stomatal response of cotton to water stress and abscisic acid as affected by water stress history. *Plant Physiol.* 65: 455-59.

Ackerson, R. C., and Krieg, D. R. 1977. Stomatal and non-stomatal regulation of water use in cotton, corn and sorghum. *Plant Physiol.* 60: 850-53.

Ackerson, R. C., Krieg, D. R., Miller, T. D., and Zartman, R. E. 1977. Water relations of field grown cotton and sorghum—temporal and diurnal changes in leaf water, osmotic and turgor potentials. *Crop Sci.* 17: 76-80.

Ackerson, R. C., Krieg, D. R., and Sung, F. J. M. 1980. Leaf conductance and osmoregulation of field-grown sorghum genotypes. *Crop Sci.* 20: 10-14.

Aharoni, N., Blumenfeld, A., and Richmond, A. E. 1977. Hormonal activity in detached lettuce leaves as affected by leaf water content. *Plant Physiol.* 59: 1169-73.

Allen, R. E., Vogel, O. E., and Peterson, C. J. 1962. Seedling emergence rate of fall-sown wheat and its association with plant height and coleoptile length. *Agron. J.* 54: 347-50.

Allerup, S. 1964. Induced transpiration changes: Effects of some growth substances added to the root medium. *Physiol. Plant.* 17: 889-908.

Al-Saadi, H., and Wiebe, H. H. 1973. Survey of the matric water of various plant groups. *Plant and Soil* 39: 253-61.

Anderson, C. H. 1971. Comparison of tillage and chemical summerfallow in a semi-arid region. *Can. J. Plant Sci.* 51: 397-403.

——. 1975. Comparison of pre-seeding tillage with total and minimal tillage by various seeding machines on spring wheat production in Southern Saskatchewan. *Can. J. Plant Sci.* 55: 59-67.

Angus, J. F., and Moncur, M. W. 1977. Water stress and phenology in wheat. *Aust. J. Agric. Res.* 28: 177-81.

Anonymous. 1975. World report on food and nutrition. *Ceres* March/April: 4. Rome: Food and Agriculture Organization of the United Nations.

Appu, P. C. 1974. The bamboo tube well: a low cost device for exploiting groundwater. *Economic and Political Weekly* 19: No. 26.

Arkin, G. F., Thompson, M., and Chaison, R. 1976. A rain-out shelter installation for studying drought stress in sorghum. *Agron. J.* 68: 429-31.

Arkin, G. F., Vanderlip, R. L., and Ritchie, J. T. 1976. A dynamic grain sorghum growth model for calculating the daily growth and development of an average grain sorghum plant in a field stand. *Trans. Amer. Soc. Agric. Eng.* 19: 622-26, 630.

Arkin, G. F., Ritchie, J. T., and Maas, S. J. 1978. Model for calculating light interception by a grain sorghum canopy. *Trans. Amer. Soc. Agric. Eng.* 21: 303-8.

Arnon, I. 1972. *Crop production in dry regions. Vol. 1. Background and principles.* London: Leonard Hill Books.

Ashraf, C. M., and Abu-Shakra, S. 1978. Wheat germination under low temperature and moisture stress. *Agron. J.* 70: 135-39.

Aslyng, H. C. 1963. Soil physics terminology. *Int. Soc. Soil Sci. Bull.* 23: 1-4.

Aung, L. H. 1974. Root-shoot relationships. In *The plant root and its environment.* ed. E. W. Carson, pp. 29-62, Charlottesville: University Press of Virginia.

Austin, R. B. 1978. Actual and potential yields of wheat and barley in the United Kingdom. *ADAS Quarterly Rev.* 29: 76-87.

Austin, R. B., Edrich, J. A., Ford, M. A., and Blackwell, R. D. 1977. The fate of dry matter, carbohydrates and ^{14}C lost from the leaves and stems of wheat during grain filling. *Ann. Bot.* 41: 1309-21.

Austin, R. B., Bingham, J., Blackwell, R. D., Evans. L. T., Ford, M. A., Morgan, C. L., and Taylor, M. 1980. Genetic improvements in winter wheat yields since 1900 and associated physiological changes. *J. Agric. Sci.* 94: 675-89.

Avery, T. E. 1968. *Interpretation of aerial photographs.* Minneapolis: Burgess Publishing Co.

Baldry, C. 1973. Progrès recents concernant l'étude du système racinaine du ble (*Triticum* spp.). *Ann. Agron.* 149: 250-76.

Bandurski, R. S. 1979. Chemistry and physiology of conjugates of indole-3-acetic acid. In *Plant growth substances*, ACS Symposium Ser. No. III, ed. N. B. Mandava, pp. 1-17. Washington: American Chemical Society.

Bange, G. G. J. 1953. On the quantitative explanation of stomatal transpiration. *Acta Bot. Neerl.* 2: 225-97.

Barnett, N. M., and Naylor, A. W. 1966. Amino acid and protein metabolism in Bermuda grass during water stress. *Plant Physiol.* 41: 1222-30.

Barrs, H. D. 1968. Determination of water deficits in plant tissues. In *Water deficits and plant growth*, ed. T. T. Kozlowski, pp. 235-368, Vol. 1. New York: Academic Press.

Barrs, H. D., and Weatherley, P. E. 1962. A re-examination of the relative turgidity technique for estimating water deficits in leaves. *Aust. J. Biol. Sci.* 15: 413-28.

Bartholic, J. T., Namkin, L. N., and Wiegand, C. L. 1972. Aerial thermal scanner to determine temperatures of soils and of crop canopies differing in water stress. *Agron. J.* 64: 603-8.

Barton Worthington, E. 1976. *Arid land irrigation in developing countries. Environmental problems and effects.* Oxford: Pergamon Press.

Beadle, C. L., Stevenson, K. R., Neumann, H. H., Thurtell, G. W., and King, K. M. 1973. Diffusive resistance, transpiration, and photosynthesis in single leaves of corn and sorghum in relation to leaf water potential. *Can. J. Plant Sci.* 53: 537-44.

Beardsell, M. F., and Cohen, D. 1975. Relationships between leaf water status, abscisic acid levels, and stomatal resistance in maize and sorghum. *Plant Physiol.* 56: 207-12.

Begg, J. E., and Turner, N. C. 1976. Crop water deficits. *Adv. Agron.* 28: 161-217.

Benasher, J. 1979. Trickle irrigation timing and its effect on plant and soil-water status. *Agric. Water Manage.* 2: 225-32.

Bengston, C., Falk, S. O., and Larsson, S. 1979. Effects of kinetin on transport rate and abscisic acid content of water stressed young wheat plants. Physiol. Plant. 45: 183-88.

Ben-Yehoshua, S., and Aloni, B. 1974. Effect of water stress on ethylene levels by detached leaves of Valencia orange (*Citrus sinensis* Osbeck). *Plant Physiol.* 53: 863–65.

Bernal, J. D. 1965. The structure of water and its biological implications. In *The state and movement of water in living organisms*. Symp. Soc. Exp. Biol. 19: 17–32, London: Cambridge University Press.

Bertrand, A. R. 1965. Water conservation through improved practices. In *Plant environment and efficient water use*, ed. W. H. Pierre, D. Kirkham, J. Pesek, and R. Shaw, pp. 207–35, Madison, Wisconsin: American Society of Agronomy–Soil Science Society of America.

Bewley, J. D. 1973. Desiccation and protein synthesis in the mass *Tortula ruralis*. *Can. J. Bot.* 51: 203–6.

——. 1974. Protein synthesis and polyribosome stability upon desiccation of the aquatic moss *Hygrohypnum luridum*. *Can. J. Bot.* 52: 423–27.

——. 1979. Physiological aspects of desiccation tolerance. *Ann. Rev. Plant Physiol.* 30: 195–238.

Beyer, E. M. 1973. Abscission: Support for the role of ethylene modification of auxin transport. *Plant Physiol.* 52: 1–5.

Beyer, E. M., and Morgan, P. W. 1971. Abscission: The role of ethylene modification of auxin transport. *Plant Physiol.* 48: 208–12.

Bhan, S., Singh, H. G., and Singh, A. 1973. Root development as an index of drought resistance in sorghum (*Sorghum bicolor* L. Moench). *Indian J. Agric. Sci.* 43: 828–30.

Bhatt, K. C., Vaishnav, P. P., Singh, Y. D., and Chinoy, J. J. 1976. Reversal of gibberellic acid induced inhibition of root growth by manganese. *Biochem. Physiol. Pflanz.* 170: 453–55.

Bidinger, F., Musgrave, R. B., and Fischer, R. A. 1977. Contribution of stored pre-anthesis assimilate to grain yield in wheat and barley. *Nature* 270: 431–33.

Bjorkman, O., Pearcy, R. W., Harrison, A. T., and Mooney, H. 1972. Photosynthetic adaptation to high temperatures: a field study in Death Valley, California. *Science* 175: 786–89.

Blad, B. L., Gardner, B. R., Watts, D. G., and Maurer, R. 1978. Crop temperature–plant water status relationships of corn and sorghum under several soil-moisture regimes. *Bull. Amer. Meteorol. Soc.* 59: 1510.

Blake, J., and Ferrell, W. K. 1977. The association between soil and xylem water potential, leaf resistance, and abscisic acid content in droughted seedlings of Douglas fir (*Pseudotsuga menziesii*). *Physiol. Plant.* 39: 106-9.

Blum, A. 1972. Effect of planting date on water use and its efficiency in dry land grain *Sorghum bicolor. Agron. J.* 64: 775-78.

——. 1974. Genotypic responses in sorghum to drought stress. 1. Response to soil moisture stress. *Crop Sci.* 14: 361-64.

——. 1975a. Infrared photography for selection of dehydration-avoidant sorghum genotypes. *Z. Pflanzenzucht.* 75: 339-45.

——. 1975b. Effect of BM gene on epicuticular wax and water relations of *Sorghum bicolor* L. Moench. *Israel J. Bot.* 24: 50-51.

——. 1979. Genetic improvement of drought resistance in crop plants: A case for sorghum. In *Stress physiology in crop plants*, ed. H. Mussell and R. C. Staples, pp. 429-45, New York: Academic Press.

Blum, A., and Sullivan, C. Y. 1972. A laboratory method for monitoring net photosynthesis in leaf segments under controlled water stress. Experiments with sorghum. *Photosynthetica* 6: 18-23.

——. 1974. Leaf water potential and stomatal activity in sorghum as influenced by soil moisture stress. *Israel J. Bot.* 23: 14-19.

Blum, A., and Ebercon, A. 1976. Genotypic responses in sorghum to drought stress. 3. Free proline accumulation and drought resistance. *Crop Sci.* 16: 428-31.

Blum, A., and Naveh, M. 1976. Improved water use efficiency in dry land grain sorghum by promoted plant competition. *Agron. J.* 68: 111-16.

Blum, A., Arkin, G. F., and Jordan, W. R. 1977. Sorghum root morphogenesis and growth. 1. Effect of maturity genes. *Crop Sci.* 17: 149-53.

Blum, A., Shertz, K. F., Toler, R. W., Welch, R. I., and Rosenow, D. T. 1978. Selection for drought avoidance in *Sorghum bicolor* using aerial photography. *Agron. J.* 70: 472-77.

Bodman, G. B., and Coleman, E. A. 1944. Moisture and energy conditions during downward entry of water into soils. *Soil Sci. Soc. Amer. Proc.* 8: 116-22.

Bolt, G. H., and Frissel, M. J. 1960. Thermodynamics of soil moisture. *Neth. J. Agric. Sci.* 8: 57-78.

Borlaug, N. 1968. The accelerated crop improvement and production programs and the agricultural revolution: Suggestions for future development. Centro Internacional de Mejoramiento de Maiz y Trigo. Mexico.

Bourque, D. P., and Naylor, A. W. 1971. Large effects of small water deficits on chlorophyll accumulation and ribonucleic acid synthesis in etiolated leaves of Jack bean (*Canavalia ensiformis* L. DC.) *Plant Physiol.* 47: 591-94.

Boyer, J. S. 1968. Relationship of water potential to growth of leaves. *Plant Physiol.* 43: 1056-62.

——. 1970a. Leaf enlargement and metabolic rates in corn, soybean and sunflower at various leaf water potentials. *Plant Physiol.* 46: 233-35.

——. 1970b. Differing sensitivity of photosynthesis to low leaf water potentials in corn and soybean. *Plant Physiol.* 46: 236-39.

Boyer, J. S., and McPherson, H. G. 1975. Physiology of water deficits in cereal crops. *Adv. Agron.* 27: 1-23.

Brady, C. J., Scott, N. S., and Munns, R. 1974. The interaction of water stress with the senescence pattern of leaves. *R. Soc. N.Z. Bull.* 12: 403-9.

Brevedon, E. R., and Hodges, H. F. 1973. Effect of moisture deficits on ^{14}C translocation in corn (*Zea mays* L.). *Plant Physiol.* 52: 436-39.

Brouwer, R. 1953. Water absorption by the roots of *Vicia faba* at various transpiration strengths. II. Causal relation between suction tension, resistance and uptake. *Proc. Kon. Med. Akad. Wet.* (c) 56: 129-36.

Brown, K. W., Jordan, W. R., and Thomas, J. C. 1976. Water stress induced alterations of the stomatal response to decreases in leaf water potential. *Physiol. Plant.* 37: 1-5.

Buckingham, E. 1907. Studies on the movement of soil moisture. *U.S. Dept. Agric., Bur. Soils Bull.* 38.

Bull, T. A. 1968. Expansion of leaf area per plant in field bean (*Vicia faba* L.) as related to daily maximum temperature. *J. Appl. Ecol.* 5: 61-68.

Bur, R., Morard, P., and Berducou, J. 1977. Importance of seminal and adventitious roots for growth and cationic nutrition of sorghum (*Sorghum dochna*). *Plant and Soil* 47: 1-12.

Burke, M. J., Gusta, L. V., Quamme, H. A., Weiser, C. J., and Li, P. H. 1976. Freezing injury in plants. *Ann. Rev. Plant. Physiol.* 27: 507-28.

Bush, T. F., Ulaby, F. T. 1978. An evaluation of radar as a crop classifier. *Remote Sens. Environ.* 7: 15-36.

Chaney, W. R., and Kozlowski, T. T. 1971. Water transport in relation to expansion and contraction of leaves and fruits of Calamondin orange. *J. Hort. Sci.* 46: 71-78.

Chatterjee, S. K., and Leopold, A. C. 1964. Kinetin and gibberellin actions on abscission processes. *Plant Physiol.* 39: 334-37.

Chatterton, N. J., Hanna, W. W., Powell, J. B., and Lee, D. R. 1975. Photosynthesis and transpiration of bloom and bloomless sorghum. *Can. J. Plant Sci.* 55: 641-43.

Childs, E. C. 1957. The anisotropic hydraulic conductivity of soil. *J. Soil Sci.* 8: 42-47.

Childs, E. C., and Collis-George, N. 1950. The permeability of porous materials. *Proc. Roy. Soc.* A201: 392-405.

Chinnici, M. F., and Peterson, D. M. 1979. Temperature and drought effects on blast and other characteristics in developing oats. *Crop Sci.* 19: 893-97.

Chotib, A., Evenson, J. P., and Harty, R. L. 1976. The role of the primary seminal root system in the promotion of normal growth in hybrid sorghum. *Seed Sci. Technol.* 4: 239-43.

Chowdury, S. I., and Wardlaw, I. F. 1978. The effect of temperature on kernel development in cereals wheat, rice and sorghum. *Aust. J. Agric. Res.* 29: 205-23.

Chrominski, A. 1972. Effect of growth regulators on water metabolism in plants part 2. *Zesz. Nauk. Uniw. Mikolaja Kopernika Toruniu Neuki Mat-Prz* 15: 125-43.

Chu, A. C. R., and Kerr, J. P. 1977. Leaf water potential and leaf extension in a sudax crop. *N.Z.J. Agric. Res.* 20: 467-70.

CIMMYT (International Maize and Wheat Improvement Centre). 1977. Report on wheat improvement. El Batan, Mexico: CIMMYT.

Clarke, J. M. 1979. Intra-plant variation in number of seeds per pod and seed weight in *Brassica napus* cv. Tower. *Can. J. Plant Sci.* 59: 959-62.

Clarke, J. M., and Elliott, C. R. 1974. Time of floral initiation in *Bromus* spp. *Can. J. Plant Sci.* 54: 475-77.

Clarke, J. M., and Simpson, G. M. 1978. Influence of irrigation and seeding rates on yield and yield components of *Brassica napus* cv. Tower. *Can. J. Plant Sci.* 58: 731-37.

Colbaugh, P. F., and Endo, R. M. 1974. Drought stress an important factor stimulating the development of *Helminthosporium sativum* on Kentucky blue grass. In *Proc. Second Int. Turfgrass Res. Conf. Blacksburg, Virginia. 1973*, ed. E. C. Roberts, pp. 328-34, Madison, Wisconsin: Amer. Soc. Agron. and Crop Sci. Soc. Amer. Inc.

Collins, J. C., and Kerrigan, A. P. 1974. The effect of kinetin and abscisic acid on water and ion transport in isolated maize roots. *New Phytol.* 73: 309-14.

Collins, W. 1978. Remote sensing of crop type and maturity—an airborne spectro radiometer was employed to detect a red spectral shift in the chlorophyll absorption edge, wheat and grain sorghum. *Photogramm. Eng. Remote Sens.* 44: 43-55.

Constable, G. A., and Hearn, A. B. 1978. Agronomic and physiological responses of soybean and sorghum crops to water deficits. 1. Growth, development and yield. *Aust. J. Plant. Physiol.* 5: 159-67.

Cook, C. W. A. 1943. A study of the roots of *Bromus inermis* in relation to drought resistance. *Ecology* 24: 161-82.

Cooper, M. J., Digby, J., and Cooper, P. J. 1972. Effects of plant hormones on the stomata of barley. A study of the interaction between abscisic acid and kinetin. *Planta* 105: 43-9.

Cowan, I. R. 1965. Transport of water in the soil-plant-atmosphere system. *J. Appl. Ecol.* 2: 221-39.

Crafts, A. S. 1968. Water structure and water in the plant body. In *Water deficits and plant growth v. I*, ed. T. T. Kozlowski, pp. 22-47, New York, San Francisco, London: Academic Press.

Craker, L. E., Chadwick, A. V., and Leather, G. R. 1970. Abscission, movement and conjugation of auxin. *Plant Physiol.* 46: 790-93.

Cram, W. J., and Pitman, M. G. 1972. The action of abscisic acid on ion uptake and water flow in plant roots. *Aust. J. Biol. Sci.* 25: 1125-32.

Cruz-Romero, G., and Ramos, C. 1979. Soil water stress and air humidity effects on the root system of sorghum. In *The soil–root interface*, ed. J. L. Harley and R. Scott-Russell, pp. 419-20, London: Academic Press.

Cummins, W. R. 1973. The metabolism of abscisic acid in relation to its reversible action on stomata in leaves of *Hordeum vulgare* L. *Planta* 114: 159-67.

Cutler, J. M., and Rains, D. W. 1978. Effects of water stress and hardening on the internal water relations and osmotic constituents of cotton leaves. *Physiol. Plant.* 42: 261-68.

Dainty, J. 1963. Water relations of plant cells. *Adv. Bot. Res.* 1: 279-326.

Darbyshire, B. 1971. Changes in indoleacetic acid oxidase activity associated with plant water potential. *Physiol. Plant.* 25: 80-4.

Das, P. K. 1972. *The monsoons.* New York: St. Martin's Press.

Das, U. S. R., and Raghavendra, A. S. 1979. Anti-transpirants for improvement of water-use efficiency of crops. *Outlook on Agriculture* 10: 92-8.

Davenport, T. L., Morgan, P. W., and Jordan, W. R. 1977. Auxin transport as related to leaf abscission during water stress in cotton. *Plant Physiol.* 59: 554-57.

——. 1980. Reduction of auxin transport capacity with age and internal water deficits in cotton petioles. *Plant. Physiol.* 65: 1023-25.

Davies, W. J. 1977. Stomatal responses to water stress and light in plants grown in controlled environments and in the field. *Crop Sci.* 17: 733-40.

——. 1978. Some effects of abscisic acid and water stress on stomata of *Vicia faba* L. *J. Exp. Bot.* 29: 175-82.

Day, W., Legg, B. J., French, B. K., Johnston, A. E., Lawlor, D. W., and Jeffers, W. D. C. 1978. Drought experiment using mobile shelters—effect of drought on barley yield, water use and nutrient uptake. *J. Agric. Sci.* 91: 599-623.

Dedio, W. 1975. Water relations in wheat leaves as screening tests for drought resistance. *Can. J. Plant Sci.* 55: 369-78.

Dedio, W., Stewart, D. W., and Green, D. G. 1976. Evaluation of photosynthesis measuring methods as possible screening techniques for drought resistance in wheat. *Can. J. Plant Sci.* 56: 243-47.

Denmead, O. T. 1973. Relative significance of soil and plant evaporation in estimating evapotranspiration. In *Plant response to climatic factors*, ed. R. O. Slatyer, pp. 505-11, Proc. Uppsala Symp. 1970, Paris: UNESCO.

Denmead, O. T., and Shaw, R. H. 1962. Availability of soil water to plants as affected by soil moisture content and meteorological conditions. *Agron. J.* 54: 385-89.

Dennett, M. D., Auld, B. A., and Elston, J. 1978. A description of leaf growth in *Vicia faba* L. *Ann. Bot.* 42: 223-32.

Dennett, M. D., Milford, J. R., and Elston, J. 1978. The effect of temperature on the relative leaf growth rate of crops of *Vicia faba* L. *Agric. Meteorol.* 19: 505-14.

Derera, N. F., Marshall, D. R., and Balaam, N. F. 1969. Genetic variability in root development in relation to drought tolerance in spring wheats. *Expl. Agric.* 5: 327-37.

Dhindsa, R. S., and Cleland, R. E. 1975. Water stress and protein synthesis. II. Interaction between water stress, hydrostatic pressure, and abscisic acid on the pattern of protein synthesis in *Avena* coleoptiles. *Plant Physiol.* 55: 782-85.

Dittrich, P., and Raschke, K. 1977. Malate metabolism in isolated epidermis of *Commelina communis* L. in relation to stomata functioning. *Planta* 134: 77-81.

Dixon, H. H. 1914. *Transpiration and the ascent of sap in plants.* London: MacMillan.

Doggett, H. 1970. *Sorghum.* London: Longmans, Green and Co.

Donald, C. M. 1968. The breeding of crop ideotypes. *Euphytica* 17: 385-403.

Dörffling, K., Streich, J., Kruse, W., and Muxfeldt, B. 1977. Abscisic acid and the after-effect of water stress on stomatal opening potential. *Z. Pflanzenphysiol.* 81: 43-56.

Doss, B. D., and Ensminger, L. E. 1958. Effect of nitrogen fertilizer on water use efficiency. *Am. Soc. Agron. 50th Ann. Mtg.* Div. VI. p. 36.

Dubbe, D. R., Farquhar, G. D., and Raschke, K. 1978. Effect of abscisic acid on the gain of the feedback loop involving carbon dioxide and stomata. *Plant Physiol.* 62: 413-17.

Duley, F. L. 1952. Relationship between surface cover and water penetration, runoff and soil losses. *Proc. 6th Int. Grassl. Congr.* 2: 942-46.

Duncan, W. G. 1967. Model building in photosynthesis. in *Harvesting the sun,* ed. A. San Pietro, F. A. Greer, and J. T. Army, New York: Academic Press.

Dungan, G. H., Lang, A. L., and Pendleton, J. W. 1958. Corn plant population in relation to soil productivity. *Adv. Agron.* 10: 435-73.

Dunlap, J. R., and Morgan, P. W. 1978. Changes in phytohormone levels during development of an early and late maturity genotype of *Sorghum bicolor* L. Moench. *Plant Physiol.* 61: Iss. 4, p. 112.

Durley, R. C., Kannangara, T., and Simpson, G. M. 1978. Analysis of abscisins and 3-indolyl acetic acid in leaves of *Sorghum bicolor* by high performance liquid chromatography. *Can. J. Bot.* 56: 157-61.

—. 1981. Leaf Analysis for abscisic-, phaseic-, and indole-3-acetic acids by HPLC. *J. Chromat.* (Submitted.)

Eastin, J. D., Taylor, A. O., and Brooking, I. 1975. Temperature influence on sorghum development and yield components. In *The physiology and yield and management of sorghum in relation to genetic improvement,* pp. 126-

47, Ann. Rept. No. 8 Univ. Nebraska, ARS/USDA Rockefeller Foundation. Lincoln, Nebraska.

Ebercon, A., Blum, A., and Jordan, W. R. 1977. Rapid colorimetric method for epicuticular wax content of sorghum leaves. *Crop Sci.* 17: 179-80.

Eck, H. V., and Musick, J. T. 1979a. Plant water stress effects on irrigated grain sorghum *Sorghum bicolor* 1. Effects on yield. *Crop Sci.* 19: 589-92.

———. 1979b. Plant water stress effects on irrigated grain sorghum *Sorghum bicolor* 2. Effects on nutrients in plant tissues. *Crop Sci.* 19: 592-98.

Edlefsen, N. E. 1941. Some thermodynamic aspects of the use of soil moisture by plants. *Trans. Amer. Geophys. Union* 22: 917-40.

Ehrler, W. L., Idso, S. B., Jackson, R. D., and Reginato, R. J. 1978. Wheat canopy temperature: Relation to plant water potential. *Agron J.* 70: 251-56.

Ehleringer, J., Bjorkman, O., and Mooney, H. A. 1976. Leaf pubescence: Effects on absorptance and photosynthesis in a desert shrub. *Science* 192: 376-77.

El-Beltagy, A. S., and Hell, M. A. 1974. Effect of water stress upon endogenous ethylene levels in *Vicia faba. New Phytol.* 73: 47-59.

El Nadi, A. H. 1969. Water relations of beans. 1. Effects of water stress on growth and flowering. *Exp. Agric.* 5: 195-207.

El Nadi, A. H., Brouwer, R., and Locher, J. T. 1969. Some responses of the root and the shoot of *Vicia faba* to water stress. *Neth. J. Agric. Sci.* 17: 133-42.

El-Sharkawi, H. M., and Springuel, I. 1977. Germination of some crop plant seeds under reduced water potential. *Seed Sci. Technol.* 5: 677-88.

El-Sharkawy, M. A., and Hesketh, J. D. 1964. Effects of temperature and water deficit on leaf photosynthetic rates of different species. *Crop Sci.* 4: 514-18.

Elston, J. F., Karamanes, A. J., Kassam, A. H., and Wadsworth, R. M., 1976. The water relations of the field bean crop. *Phil. Trans. Roy. Soc. Lond.* B273: 581-91.

Erdei, L., Toth, I., and Zsoldos, F. 1979. Hormonal regulation of Ca^{2+}-stimulated K^+ influx and Ca^{2+}, K^+-ATPase in rice roots: *In vivo* and *in vitro* effects of auxins and reconstitution of the ATPase. *Physiol. Plant.* 45: 448-52.

Erickson, P. I., Kirkham, M. B., and Stone, J. F. 1979. Growth, water relations and yield of wheat planted in four row directions. *Soil Sci. Soc. Am. J.* 43: 570-74.

Eslick, R. F., and Hockett, E. A. 1974. Genetic engineering as a key to water-use efficiency. *Agric. Meteorol.* 14: 13–23.

Evans, L. E., and Bhatt, G. M. 1977. Influence of seed size, protein content and cultivar on early seedling vigor in wheat. *Can. J. Plant. Sci.* 57: 929–35.

Evans, L. T. 1963. *Environmental control of plant growth.* New York: Academic Press.

Evans, L. T., and Dunstone, R. L. 1970. Some physiological aspects of evolution in wheat. *Aust. J. Biol. Sci.* 23: 725–41.

Farbrother, H. G. 1957. On an electrical resistance technique for the study of soil moisture problems in the field. *Emp. Cotton Growing Rev.* 34: 1–22.

Felger, R. S., and Nabham, G. P. 1976. Deceptive barreness. *Ceres* March/April: 34–39. Rome: Food and Agriculture Organization of the United Nations.

Fenton, R., Davies, W. J., and Mansfield, T. A. 1977. The role of farnesol as a regulator of stomatal opening in *Sorghum bicolor. J. Exp. Bot.* 28: 1043–53.

Fereres, E., Acevado, E., Henderson, D. W., and Hsiao, T. C. 1976. Water potential components and growth of sorghum in the field under water stress. *Plant Physiol.* 57: Suppl. p. 44.

Fereres, E., Acevado, E., Henderson, D. W., and Hsiao, T. C. 1978. Seasonal changes in water potential and turgor maintenance in sorghum and maize under water stress. *Physiol. Plant.* 44: 261–67.

Filatov, P. A., and Frolova, R. I. 1972. Increasing drought resistance of sunflower plants by pre-sowing hardening of seeds. *Sov. Plant Physiol.* 19: 660–64.

——. 1974. Use of pre-sowing hardening of sunflower seeds for seed production. *Fiziol. Rast.* 21: 187–91.

Finch-Savage, W. E., and Elston, J. F. 1976. The death of leaves in crops of field beans. *Ann. Appl. Biol.* 85: 463–65.

Fischer, R. A. 1968. Stomatal opening: role of potassium uptake by guard cells. *Science* 160: 784–85.

——. 1973. The effect of water stress at various stages of development on yield processes in wheat. In *Plant response to climatic factors,* ed. R. O. Slatyer, pp. 233–41, Proc. Uppsala Symp. 1970, Paris: UNESCO.

Fischer, R. A., and Kohn, G. D. 1966a. The relationship between evapotranspiration and growth in the wheat crop. *Aust. J. Agric. Res.* 17: 255–67.

——. 1966b. The relationship of grain yield to vegetative growth and post-flowering leaf area in the wheat crop under conditions of limited soil moisture. *Aust. J. Agric. Res.* 17: 281–95.

Fischer, R. A. and Maurer, R. 1978. Drought resistance in spring wheat cultivars. 1. Grain yield responses. *Aust. J. Agric. Res.* 29: 897–912.

Flowers, T. J., Troke, P. F., and Yeo, A. R. 1977. The mechanism of salt tolerance in halophytes. *Ann. Rev. Plant. Physiol.* 28: 89–121.

Frank. A. B., and Willis, W. O. 1972. Influence of wind breaks on leaf water status in spring wheat. *Crop Sci.* 12: 668–72.

Frank, A. B., Power, J. F., and Willis, W. O. 1973. Effect of temperature and plant water stress on photosynthesis, diffusion resistance, and leaf water potential in spring wheat. *Agron. J.* 65: 777–80.

Freeman, B. M., Blackwell, J., and Garzoli, K. V. 1976. Irrigation frequency and total water application with trickle and furrow systems. *Agric. Water Manage.* 1: 21–32.

French, R. J. 1978. The effect of fallowing on the yield of wheat. 1. The effect on soil water storage and nitrate supply. *Aust. J. Agric. Res.* 29: 653–58.

Fuchs, M., Stanhill, G., and Moreshet, S. 1976. Effect of increasing foliage and soil reflectivity on the solar radiation balance of wide-row grain sorghum. *Agron. J.* 68: 865–71.

Gacso, L., and Lassanyi, Z. 1978. New anti-transpirants—substituted alpha-hydroxysulfonates. 2. Their effect on the water metabolism of plants. Botanikai Kozlemenyek (Botanical Publications) Vol. 65: 239.

Gaff, D. F. 1971. Desiccation-tolerant flowering plants in Southern Africa. *Science* 174: 1033–34.

Gallacher, A. G., and Sprent, J. I. 1978. The effect of different water regimes on growth, and nodule development of greenhouse-grown *Vicia faba. J. Exp. Bot.* 29: 413–23.

Gallagher, J. N., Biscoe, P. V., and Hunter, B. 1976. Effect of drought on grain growth. *Nature* 264: 541–42.

Galtung, J. 1979. Development, environment and technology. Towards a technology for self reliance. United Nations Conference on Trade and Technology. New York: United Nations.

Gardner, W. R. 1964. Relation of root distribution to water uptake and availability. *Agron J.* 56: 41–45.

——. 1973. Internal water status and plant response in relation to the external water regime. In *Plant response to climatic factors*, ed. R. O. Slatyer, pp. 221–25, Proc. Uppsala Symp. 1970, Paris: UNESCO.

Gardner, W. R., and Nieman, R. H. 1964. Lower limit of water availability to plants. *Science* 143: 1460–62.

Gardner, W. R., and Ehlig, C. F. 1965. Physical aspects of the internal water relations of plant leaves. *Plant Physiol.* 40: 705-10.

Garduño, M. A. 1977. Arms for the struggle. *Ceres* March/April: 41-44. Rome: Food and Agriculture Organization of the United Nations.

Gates, C. T. 1957. The response of the young tomato plant to a brief period of water shortage. 3. Drifts in nitrogen and phosphorus. *Aust. J. Biol. Sci.* 10: 125-46.

——. 1968. Water deficits and growth of herbaceous plants. In *Water deficits and plant growth,* ed. T. T. Kozlowski, pp. 135-90, Vol. 2, New York: Academic Press.

Gates, D. M., Alderfer, R., and Taylor, E. 1968. Leaf temperature of desert plants. *Science* 159: 994-95.

Genkel, P. A., Badanova, K. A., Prusakova, L. D., and Bokarev, K. S. 1979. Multiple method to increase the heat and drought resistance of spring wheat. *Sov. Plant Physiol.* 26: 518-23.

Gibson, P. T., and Schertz, K. F. 1977. Growth analysis of a *Sorghum bicolor* hybrid and its parents. *Crop Sci.* 17: 387-91.

Giles, K. L., Cohen, D., and Beardsall, M. F. 1976. Effects of water stress on ultrastructure of leaf cells of *Sorghum bicolor. Plant Physiol.* 57: 11-14.

Giordano, R., Salleo, A., Salleo, S., and Wanderlingh, F. 1978. Flow in xylem vessels and Poiseuille's law. *Can. J. Bot.* 56: 333-38.

Glinka, Z. 1973. Abscisic acid effect on root exudation related to increased permeability to water. *Plant Physiol.* 51: 217-19.

——. 1977. Effect of abscisic acid and of hydrostatic pressure gradient on water movement through excised sunflower roots. *Plant Physiol.* 59: 933-35.

——. 1980. Abscisic acid promotes both volume and flow and ion release to the xylem in sunflower roots. *Plant Physiol.* 65: 537-40.

Glover, J. 1959. The apparent behaviour of maize and sorghum stomata during and after drought. *J. Agric. Sci.* 53: 412-16.

Goplen, B. P., Baenziger, H., Bailey, L. D., Gross, A. T. H., Hanna, M. R., Michaud, R., Richards, K. W., and Waddington, J. 1980. *Growing and managing alfalfa in Canada.* Agriculture Canada Publ. 1705. Ottawa.

Government of India. 1974. *Crop production strategy in rainfed areas under different weather conditions during 1974-75.* Indian Council of Agric. Res., Agric. Universities and Dryland Agric. Res. Centres. New Delhi, India.

Govinden, J. R. N. 1977. The effect of abscisic acid sprays on aspects of the growth and productivity of water-stressed wheat (*Triticum aestivum* L). M.Sc. Thesis. University of Saskatchewan, Saskatoon, Canada.

Green, P. B. 1968. Growth physics in Nitella: A method for continuous *in vivo* analysis of extensibility based on a micro-manometer technique for turgor pressure. *Plant Physiol.* 43: 1169-84.

Grenot, C. J. 1974. Physical and vegetational aspects of the Sahara desert. In *Desert biology Vol. 2*, ed. G. W. Brown, New York: Academic Press.

Grigg, D. B. 1974. *The agricultural systems of the world. An evolutionary approach.* Cambridge: Cambridge University Press.

Gul, A., and Allan, R. E. 1976a. Stand establishment of wheat lines under different levels of water potential. *Crop Sci.* 16: 611-15.

——. 1976b. Interrelationships of seedling vigor criteria of wheat under different field situations and soil water potentials. *Crop Sci.* 16: 615-18.

Gurr, C. G., Marshall, T. J., and Hutton, J. T. 1952. Movement of water in soil due to a temperature gradient. *Soil Sci.* 74: 335-45.

Haas, H. J., and Willis, W. 0. 1968. Conservation bench terraces in North Dakota. *Trans. ASAE* 11: 396-98.

Haines, F. M. 1950. The relations between cell dimensions, osmotic pressure and turgor pressure. *Ann. Bot.* 14: 385-94.

Hall, A. E., and Kaufman, M. R. 1975. Regulation of water transport in the soil-plant-atmosphere continuum. In *Perspectives of biophysical ecology*, ed. D. M. Gates and R. B. Schmerl, pp. 187-202, Berlin: Springer-Verlag.

Hall, M. A., Kapuya, J. A., Sivakumaran, S., and John, A. 1977. The role of ethylene in the response of plants to stress. *Pestic. Sci.* 8: 217-23.

Hanks, R. J. 1974. Model for predicting plant yield as influenced by water use. *Agron. J.* 66: 660-65.

Harlan, J. R. 1956. *Theory and dynamics of grassland agriculture.* Princeton, N.J.: Van Nostrand Co.

Harrison, M. A., and Walton, D. C. 1975. Abscisic acid metabolism in water stressed bean leaves. *Plant Physiol.* 56: 250-54.

Hartsock, T. L., and Nobel, P. S. 1976. Watering converts a CAM plant to daytime CO_2 uptake. *Nature* 262: 574-76.

Hartt, C. E. 1967. Effect of moisture supply upon translocation and storage of ^{14}C in sugar cane. *Plant Physiol.* 42: 338-46.

Hatfield, J. L., Hipps, L. E., and Walker, G. K. 1978. Microclimate variations between grain-sorghum and soybean canopies and the effects on growth and yield. *Bull. Amer. Meteorol. Soc.* 59: 1509-10.

Hawkes, J. G., Williams, J. T., and Hanson, J. 1976. *A bibliography of plant genetic resources.* Rome: International Board for Plant Genetic Resources.

Heilmann, B., Hartung, W., and Gimmler, H. 1980. The distribution of abscisic acid between chloroplasts and cytoplasm of leaf cells and the permeability of the chloroplast envelope for abscisic acid. *Z. Pflanzenphysiol.* 97: 67-78.

Heilman, J. L., Kanemasu, E. T., Rosenberg, N. J., and Blad, B. L. 1976. Thermal scanner measurement of canopy temperatures to estimate evapotranspiration. *Remote Sens. Environ.* 5: 137-45.

Helmerick, R. H., and Pfeiffer, R. P. 1954. Differential varietal responses of winter wheat germination and early growth to controlled limited moisture conditions. *Agron. J.* 46: 560-62.

Henckel, P. A. 1964. Physiology of plants under drought. *Ann. Rev. Plant Physiol.* 15: 363-86.

Henckel, P. A., and Pustovoitova, T. N. 1977. The role of growth inhibitors in adaptive modification of plants under drought conditions. In *Plant growth regulators,* Proc. Second Int. Symp. Sofia. ed. T. I. Kudrev, pp. 236-39, Sofia, Bulgaria: Bulgarian Academy of Sciences.

Henzell, R. G., McCree, K. J., van Bavel, C. H. M., and Schertz, K. F. 1975. Method for screening sorghum genotypes for stomatal sensitivity to water deficits. *Crop Sci.* 15: 516-18.

——. 1976. Sorghum genotype variation in stomatal sensitivity to leaf water deficit. *Crop Sci.* 16: 660-62.

Hewitt, J. S., and Dexter, A. R. 1979. An improved model of root growth in structured soil. *Plant and Soil* 52: 325-43.

Heyland, K-U., Solansky, S., and Aufhammer, W. 1975. The effect of CCC and gibberellic acid treatments on the yield components of spring barley. *Z. fur Acker-und Pflanzenbau* 141: 109-19.

Hiler, E. A., van Bavel, C. H. M., Hossain, M. M., and Jordan, W. R. 1972. Sensitivity of southern peas to plant water deficit at three growth stages. *Agron. J.* 64: 60-64.

Hillel, D. 1971. *Soil and water.* New York, San Francisco, London: Academic Press.

Hiron, R. W. P., and Wright, S. T. C. 1973. The role of endogenous abscisic acid in the response of plants to stress. *J. Exp. Bot.* 24: 769-81.

Ho, D. T-H., and Varner, J. E. 1974. Hormonal control of messenger ribonucleic acid metabolism in barley aleurone layers. *Proc. Natl. Acad. Sci. U.S.A.* 71: 4783-86.

Hoad, G. V. 1975. Effect of osmotic stress on abscisic acid levels in xylem sap of sunflower (*Helianthus annuus* L.). *Planta* 124: 25-9.

——. 1978. Effect of water stress on abscisic acid levels in white lupin (*Lupinus albus* L.) fruit, leaves and phloem exudate. *Planta* 142: 287-90.

Hodges, T., Kanemasu, E. T., and Teare, I. D. 1979. Modelling dry matter accumulation and yield of grain sorghum (*Sorghum bicolor*). *Can. J. Plant Sci.* 59: 803-18.

Holmgren, P., Jarvis, P. G., and Jarvis, M. S. 1965. Resistances to carbon dioxide and water vapour transfer in leaves of different plant species. *Physiol. Plant.* 18: 557-73.

Horton, R. F., and Osborne, D. J. 1967. Senescence, abscission and cellulase activity in *Phaseolus vulgaris*. *Nature* 214: 1086-88.

Hsiao, T. C. 1973. Plant resources to water stress. *Ann. Rev. Plant Physiol.* 24: 519-70.

Hsiao, T. C., and Acevedo, E. 1974. Plant response to water deficits, water use efficiency, and drought resistance. *Agric. Meteorol.* 14: 59-84.

Huck, M. G., Klepper, B., and Taylor, H. M. 1970. Diurnal variations in root diameter. *Plant Physiol.* 45: 529-30.

Huffaker, R. C., Radin, T., Kleinkopf, G. E., and Cox, E. L. 1970. Effects of mild water stress on enzymes of nitrate assimilation and of the carboxylative phase of photosynthesis in barley. *Crop. Sci.* 10: 471-74.

Hulse, J. H., Laing, E. M., and Pearson, O. E. 1980. *Sorghum and the millets: their composition and nutritive value.* New York: Academic Press.

Hultquist, J. H. 1973. *Physiologic and morphological investigations of grain sorghum* (Sorghum bicolor L. Moench). *1. Vascularization: 2. Response to internal drought stress.* Ph.D. thesis, University of Nebraska, Lincoln, Nebraska, 140 pp.

Hurd, E. A. 1964. Root study of three wheat varieties and their resistance to drought and damage by soil cracking. *Can. J. Plant Sci.* 44: 240-48.

——. 1968. Growth of roots of seven varieties of spring wheat at high and low moisture levels. *Agron. J.* 60: 201-5.

——. 1969. A method of breeding for yield of wheat in semi-arid climates. *Euphytica* 18: 217-26.

——. 1971. Can we breed for drought resistance? In *Drought injury and resistance in crops*, ed. K. L. Larson and J. D. Eastin, pp. 77–88, Crop Sci. Soc. America Spec. Pub. 2, Madison, Wisconsin.

——. 1974. Phenotype and drought tolerance in wheat. *Agric Meteorol.* 14: 39–55.

——. 1976. Plant breeding for drought resistance. In *Water deficits and plant growth*, ed. T. T. Kozlowski, pp. 317–53, Vol. 4, New York: Academic Press.

Hurd, E. A., Townley-Smith, T. F., Patterson, L. A., and Owen, C. H. 1972. Techniques used in producing Wascana wheat. *Can. J. Plant. Sci.* 52: 689–91.

Husain, I., and Aspinall, D. 1970. Water stress and apical morphogenesis in barley. *Ann. Bot.* 34: 393–407.

Incoll. L. D., and Whitelam, G. C. 1977. The effect of kinetin on stomata of the grass *Anthephora pubescens* Nees. *Planta* 137: 243–45.

International Biological Program, Matador Project. 1974. Technical reports No's 1–61. University of Saskatchewan, Saskatoon, Canada: Canadian Committee for the International Biological Program.

International Rice Research Institute. 1976. *Annual report for 1975.* Los Banos, Phillipines.

Inuyama, S. 1978a. Varietal differences in leaf water potential, leaf diffusive resistance and grain yield of grain-sorghum affected by drought stress. *Jap. J. Crop. Sci.* 47: 255–61.

——. 1978b. Effects of plant densities under two irrigation regimes on leaf water potential, leaf diffusive resistance during drought stress period and grain yield of grain sorghum. *Jap. J. Crop Sci.* 47: 596–601.

Inuyama, S., Musick, J. T., and Dusek, D. A. 1976. Effect of plant water deficits at various growth stages on growth, grain yield and leaf water potential of irrigated grain sorghum. *Proc. Crop Sci. Soc. Japan* 45: 298–307.

Iowa State University Graduate College. 1980. *Plant roots: a compilation of ten seminars.* 164 pp.

Ishag, H. M. H. 1969. *Physiology of seed yield in* Vicia faba L. Ph.D. Thesis, University of Reading.

Itai, C., and Vaadia, Y. 1965. Kinetin-like activity in root exudate of water stressed sunflower plants. *Physiol. Plant.* 18: 441–44.

——. 1971 Cytokinin activity in water-stressed plants. *Physiol. Plant.* 47: 87–90.

Itai, C., and Meidner, H. 1978. Effects of abscisic acid on solute transport in epidermal tissue. *Nature* 271: 653-54.

Jackson, M. B., and Osborne, D. J. 1970. Ethylene, the natural regulator of leaf abscission. *Nature* 225: 1019-22.

Jackson, M. B., and Campbell, D. J. 1976. Waterlogging and petiole epinasty in tomato: the role of ethylene and low oxygen. *New Phytol.* 76: 21-9.

Jacobsen, J. V. 1977. Regulation of ribonucleic acid metabolism by hormones. *Ann. Rev. Plant Physiol.* 28: 537-64.

Jensen, N. F. 1978. Limits to growth in world food production. *Science* 201: 317-20.

Jika, N. I., Saint-Pierre, C. A., and Denis, J. C. 1980. Adaptation of sorghum cultivars to different water supply conditions. *Can. J. Plant Sci.* 60: 233-39.

Johansen, S. 1954. Effect of indole acetic acid on stomata and photosynthesis. *Physiol. Plant.* 7: 531-37.

Johnson, R. R., and Moss, D. N. 1976. Effect of water stress on $^{14}CO_2$ fixation and translocation in wheat during grain filling. *Crop Sci.* 16: 697-701.

Johnson, R. R., Frey, N. M., and Moss, D. N. 1974. Effect of water stress on photosynthesis and transpiration of flag leaves and spikes of barley and wheat. *Crop Sci.* 14: 728-31.

Johnson, V. A., Shafer, S. L., and Schmidt, J. W. 1968. Regression analysis of general adaptation in hard red winter wheat (*Triticum aestivum* L.). *Crop Sci.* 8: 187-91.

Jones, F. R. 1928. Development of the bacteria causing wilt in the alfalfa plant as influenced by growth and winter injury. *J. Agric. Res.* 37: 507-69.

Jones, H. G. 1977. Aspects of water relations of spring wheat (*Triticum aestivum* L.) in response to induced drought. *J. Agric. Sci.* 88: 267-82.

——. 1979. Stomatal behaviour and breeding for drought resistance. In *Stress physiology in crop plants*, ed. H. Mussell and R. C. Staples, pp. 407-28, New York: Wiley.

Jones, L. H. 1963. The effect of soil moisture gradients on the growth and development of broad beans. *Hort. Res.* 3: 13-26.

Jones, M. M., and Turner, N. C. 1978. Osmotic adjustment in leaves of sorghum in response to water deficits. *Plant Physiol.* 61: 122-26.

Jones, M. M., and Rawson, H. M. 1979. Influence of rate of development of leaf water deficits upon photosynthesis, leaf conductance, water use efficiency and osmotic potential in sorghum. *Plant Physiol.* 45: 103-11.

Jones, M. M., Osmond, C. E., and Turner, N. C. 1980. Accumulation of solutes in leaves of sorghum and sunflower in response to water deficits. *Aust. J. Plant Physiol.* 7: 193-205.

Jordan, W. R., Morgan, P. W., and Davenport, T. L. 1972. Water stress enhances ethylene-mediated leaf abscission in cotton. *Plant Physiol.* 50: 756-58.

Jordan, W. R., Miller, F. R., and Morris, D. E. 1979. Genetic variation in root and shoot growth of sorghum in hydroponics. *Crop Sci.* 19: 468-72.

Jung, J., and Dressel, J. 1973. Investigation on the effect of growth regulators CCC, CMH and DMC on wheat and barley. *Z. Pflanzenernaehr Bodenk.* 135: 98-111.

Jury, W. A., Frenkel, H., Fluhler, H., Devitt, D., and Stolzy, L. H. 1978. Use of saline irrigation waters and minimal leaching for crop production. *Hilgardia* 46: 169-92.

Kanemasu, E. T., Stone, L. R., and Powers, W. L. 1976. Evapotranspiration model tested for soybean and sorghum. *Agron. J.* 68: 569-72.

Kannangara, T., Durley, R. C., and Simpson, G. M. 1978. High performance liquid chromatographic analysis of cytokinins in *Sorghum bicolor* leaves. *Physiol. Plant.* 44: 295-99.

Kannangara, T., Durley, R. C., Simpson, G. M., and Stout, D. G. 1981. Drought resistance of *Sorghum bicolor.* 4. Hormonal changes in relation to water deficit stress in field grown plants. *Can. J. Bot.* (Submitted.)

Karamanos, A. J. 1976. *An analysis of the effect of water stress on leaf area growth in* Vicia faba *L. in the field.* Ph.D. Thesis, University of Reading.

——. 1978a. Water stress and leaf growth of field beans (*Vicia faba* L.) in the field: Leaf number and total leaf area. Ann. Bot. 42: 1393-1402.

——. 1978b. Understanding the origin of the responses of plants to water stress by means of an equilibrium model. *Praktika Acad. Athens* 53: 308-41.

——. 1979. Water stress: A challenge for the future of agriculture. In *Plant regulation and world agriculture*, ed. T. K. Scott, pp. 415-55, New York: Plenum Publishing Corporation.

——. 1980. Response in plant water status to integrated values of soil matric potential calculated from soil water depletion by a field bean crop. *Aust. J. Plant Physiol.* 7: 51-66.

Karmoker, J. L., and Van Steveninck, R. F. M. 1978. Simulation of volume flow and ion flux by abscisic acid in excised root systems of *Phaseolus vulgaris* L. cv. Redland Pioneer. *Planta* 141: 37–43.

——. 1979. The effect of abscisic acid on the uptake and distribution of ions in intact seedlings of *Phaseolus vulgaris* cv. Redland Pioneer. *Physiol. Plant.* 45: 453–59.

Kassam, A. H. 1971. *Some physical aspects of the water relations of* Vicia faba *L.* Ph.D. Thesis, University of Reading.

——. 1972. Determination of water potential and tissue characteristics of leaves of *Vicia faba* L. *Hort. Res.* 12: 13–23.

——. 1973. Influence of light and water deficit upon diffusive resistance of leaves of *Vicia faba* L. *New Phytol.* 72: 557–70.

——. 1975. Wilting of leaves of *Vicia faba* L. *Ann. Bot.* 39: 265–71.

Kassam, A. H., and Elston, J. 1974. Seasonal changes in the status of water and tissue characteristics of leaves of *Vicia faba* L. *Ann. Bot.* 38: 419–29.

——. 1976. Changes with age in the status of water and tissue characteristics in individual leaves of *Vicia faba* L. *Ann. Bot.* 40: 669–79.

Kaul. R. 1974. Potential net photosynthesis in flag leaves of severely drought-stressed wheat cultivars and its relationship to grain yield. *Can. J. Plant Sci.* 54: 811–15.

Kaul, R., and Crowle, W. L. 1971. Relations between water status, leaf temperature, stomatal aperture and productivity of some wheat varieties. *Z. Pflanzenzuchtg.* 65: 233–43.

——. 1974. An index derived from photosynthetic parameters for predicting grain yields of drought stressed wheat cultivars. *Z. Pflanzenzuchtg.* 71: 42–51.

Kawase, M. 1972. Effect of flooding on ethylene concentration in horticultural plants. *J. Am. Soc. Hortic. Sci.* 97: 584–88.

——. 1974. Role of ethylene in induction of flooding damage in sunflower. *Physiol. Plant.* 31: 29–38.

Kazama, H., and Katsumi, M. 1978. The role of the osmotic potential of the cell in auxin induced cell elongation. *Plant Cell Physiol.* 19: 1145–50.

Keck, R. W., and Boyer, J. S. 1974. Chloroplast response to low leaf water potentials III. Differing inhibition of electron transport and photophosphorylation. *Plant Physiol.* 53: 474–79.

Kemle, H. 1965. Kinetin-like factors in the root exudate of sunflowers. *Proc. Natl. Acad. Sci. U.S.A.* 53: 1302–7.

Kilcher, M. R. 1961. *Row spacing affects yields of forage grasses in the brown soil zone of Saskatchewan.* Canada Dept. Agric. Publ. 1100, Ottawa, Canada.

Kilcher, M. R., and Heinrichs, D. H. 1958. The performance of three grasses when grown alone, in mixture with alfalfa, and in alternate rows with alfalfa. *Can. J. Plant Sci.* 38: 252-59.

King, R. W., and Evans, L. T. 1977. Inhibition of flowering in *Lolium temulentum* L. by water stress: a role for abscisic acid. *Aust. J. Plant Physiol.* 4: 225-33.

Klages, K. H. W. 1942. *Ecological crop geography.* New York: MacMillan.

Klepper, B., Taylor, H. M., Huck, M. G., and Fiscus, F. L. 1973. Water relations and growth of cotton in drying soil. *Agron. J.* 65: 307-10.

Knowles, R. P., and Buglass, E. 1966. *Crested wheat grass.* Canada Dept. Agric. Publ. 1295. Ottawa.

Koestler, A. 1967. *The ghost in the machine.* London, England: Hutchinson.

Köppen, W. 1931. *Grundriss der Klimatologie.* Berlin: De Gruyter Verlag.

Kozinka, V. 1967. Water uptake during rapid changes in transpiration induced by the presence of high concentration of growth substances in the root medium. *Biol. Plant.* 9: 222-23.

Kozlowski, T. T. 1976. Water supply and leaf shedding. In *Water deficits and plant growth*, ed. T. T. Kozlowski, Vol. 4, pp. 191-231, New York, San Francisco, London: Academic Press.

Kramer, P. J. 1969. *Plant and soil water relationships. A modern synthesis.* New York: McGraw-Hill.

Kriedemann, P. E., Loveys, B. R., and Downton, W. J. S. 1975. Internal control of stomatal physiology and photosynthesis. II. Photosynthetic responses to phaseic acid. *Aust. J. Plant Physiol.* 2: 553-67.

Krieg, D. R., Goodin, J. R., and Stevens, R. G. 1977. *Quantity and quality considerations for water use efficiency in irrigation.* Texas Tech. University and U.S. Dept. Interior, Office of Water Research and technology. Washington, D.C.

Kushnirenko, M. D., Kryukova, E. V., and Pecherskaya, S. N. 1973. Water retaining ability and proteins of leaves and chloroplasts in plants with different drought resistances. *Fiziol. Rast.* 20: 582-89.

Kutschera, L. 1960. *Wurzelatlas Mitteleuropäischer Ackerungkräuter und Kulturpflanzen*, Frankfurt a.M: DLG Verlags.

Langhans, R. W. 1978. *A growth chamber manual.* Ithaca, New York: Comstock Publishing Associates.

Langlet, A. 1973. Drought effects on growth and yield of grain sorghum. *Annales Agron.* 24: 307-38.

Larcher, W. 1975. *Physiological plant ecology.* Berlin: Springer-Verlag.

Larque-Saavedra, A., and Wain, R. L. 1976. Studies on plant growth regulating substances. 42. Abscisic acid as a genetic character related to drought tolerance. *Ann. Appl. Biol.* 83: 291-97.

Laszlo, E. 1972. *The systems view of the world.* New York: George Braziller.

Lawlor, D. W. 1969. Plant growth in polytheylene glycol solutions in relation to the osmotic potential of the root medium and the leaf water balance. *J. Exp. Bot.* 20: 895-911.

——. 1972. Growth and water use of *Lolium perenne.* I. Water transport. *J. Appl. Ecol.* 9: 79-98.

Legg, B. J., Day, W., Brown, N. J., and Smith, G. J. 1978. Small plots and automatic rain shelters a field appraisal. *J. Agric. Sci.* 91: 321-36.

Levitt, J. 1972. *Responses of plants to environmental stresses.* New York: Academic Press.

——. 1978. Crop tolerance to suboptimal land conditions a historical review. In *Crop tolerance to suboptimal land conditions,* ed. G. A. Jung, Am. Soc. Agron. Spec. Publ. 32, pp. 161-71. Madison, Wisconsin.

——. 1980a. *Responses of plants to environmental stresses. Vol. 1. Chilling, freezing and high temperature stresses.* New York: Academic Press.

——. 1980b. *Responses of plants to environmental stresses. Vol. 2. Water, radiation, salt and other stresses.* New York: Academic Press.

Lewis, M. C., and Callaghan, T. V. 1976. Tundra. In *Vegetation and the atmosphere. Vol. 2. Case studies,* ed. J. L. Monteith, pp. 399-433, New York: Academic Press.

Lewis, R. B., Hiler, E. A., and Jordan, W. R. 1974. Susceptibility of grain sorghum to water deficit at three growth stages. *Agron. J.* 66: 589-91.

Li, P. H., and Sakai, A. 1978. *Plant cold hardiness and freezing stress.* New York: Academic Press.

Liang, G. H., Dayton, A. D., Chu, C. C., and Casady, A. J. 1975. Heritability of stomatal density and distribution in leaves of grain sorghum. *Crop Sci.* 15: 567-70.

Lieth, H. 1975. Primary production of the major vegetation units of the world. In *Primary productivity of the Biosphere*, ed. H. Lieth and R. H. Whittaker, pp. 203-5, New York: Springer-Verlag.

Lipman, C. B. 1941. The successful revival of *Nostoc commune* from an herbarium specimen 87 years old. *Bull. Torrey Bot. Club* 68: 664.

Livne, A., and Vaadia, Y. 1965. Stimulation of transpiration rate in barley leaves by kinetin and gibberellic acid. *Physiol. Plant.* 39: 658-64.

——. 1972. Water deficits and hormone relations. In *Water deficits and plant growth*, ed. T. T. Kozlowski, pp. 255-75, New York: Academic Press.

Lockhart, J. A. 1965. Cell extension. In *Plant biochemistry*, ed. J. Bonner and J. E. Varner, pp. 827-49, New York: Academic Press.

Lomte, M. H., Soudge, V. D., Upadhyaya, U. C., and Varada, S. B. 1979. Leaf area constants for new varieties of sorghum (*Sorghum bicolor*). *Indian J. Agric. Sci.* 49: 392-94.

Louie, D. S., and Addicott, F. T. 1970. Applied auxin gradients and abscission in explants. *Plant Physiol.* 45: 654-57.

Loveys, B. R. 1977. The intracellular location of abscisic acid in stressed and non-stressed leaf tissue. *Physiol. Plant.* 40: 6-10.

Ludlow, M. M., and Ng, T. T. 1976. Effect of water deficit on carbon dioxide exchange and leaf elongation rate of *Panicum maximum* var. trichoglume. *Aust. J. Plant Physiol.* 3: 401-13.

Luke, H. H., and Freeman, T. E. 1967. Rapid bioassay for phytokinins based on transpiration of excised oak leaves. *Nature* 215: 874-75.

——. 1968. Stimulation of transpiration by cytokinins. *Nature* 217: 873-74.

Lyons, J. M., Graham, D., and Raison, J. K. 1980. *Low temperature stress in plants.* New York: Academic Press.

Makkink, G. F., and van Heemst, H. D. J. 1956. The actual evapotranspiration as a function of the potential evapotranspiration and the soil moisture tension. *Neth. J. Agric. Sci.* 4: 67-71.

Mali, C. V., Varade, S. B., Musande, V. G., and Chalwade, P. B. 1978. Critical soil water potential and seed hydration for germination of grain sorghum. *Curr. Sci.* 47: 587-88.

Mallory, W. H. 1926. *China, land of famine.* New York: Amer. Geog. Soc.

Mangelsdorf, P. C. 1966. Genetic potentials for increasing yields of food crops and animals. In *Prospects of world food supply.* Proc. Nat. Acad. Sci. Symposium 52: 370-75.

Mansfield, T. A. 1967. Stomatal behaviour following treatment with auxin-like substances and phenol mercuric acetate. *New Phytol.* 66: 325-30.

——. 1976a. Delay in the response of stomata to abscisic acid in CO_2-free air. *J. Exp. Bot.* 27: 559-64.

——. 1976b. Stomatal behaviour. *Phil. Trans. Roy. Soc. Lond. B.* 273: 541-50.

Mansfield, T. A., and Jones, R. J. 1971. Effects of abscisic acid on potassium uptake and starch content of stomatal guard cells. *Planta* 101: 147-58.

Mansfield, T. A., Wellburn, A. R., and Moreira, T. J. S. 1978. The role of abscisic acid and farnesol in the alleviation of water stress. *Phil. Trans. Roy. Soc. Lond. B.* 284: 471-82.

Markowski, A., Grzesiak, S., and Schramel, M. 1974. Susceptibility of six species of cultivated plants to sulfur dioxide under optimum soil moisture and drought conditions. *Bull. Acad. Pol. Sci. Ser. Sci. Biol.* 22: 889-98.

Marshall, T. J. 1959. *Relations between water and soil.* Tech. Comm. 50. Commonwealth Agricultural Bureaux, Farnham, Berkshire, England.

Martin, W. W. 1977. Selective breeding may hold the key to drought resistance in sorghum hardiness. *Agric. Res.* 26: 3-5.

Maximov, N. A. 1929. *The plant in relation to water.* London: Allen and Unwin.

Maxwell, J., and Redmann, R. E. 1974. *Water relations and growth of Agropyron dasystachyum: Recovery from soil moisture stress.* Tech. Rpt. No. 64, Matador Project (IBP), University of Saskatchewan, Saskatoon, Canada.

May, L. H., and Milthorpe, F. L. 1062. Drought resistance of crop plants. *Field Crop Abs.* 15: 171-79.

Mayaki, W. C., Stone, L. R., and Teare, I. D. 1976. Irrigated and non-irrigated soybean, corn and grain sorghum root systems. *Agron. J.* 68: 432-34.

McCauley, G. N., Stone, J. F., and Chin Choy, E. W. 1978. Evapotranspiration reduction by field geometry effects in peanuts and grain sorghum. *Agric. Meteorol.* 19: 295-304.

McCree, K. J. 1974. Changes in stomatal response characteristics of grain sorghum produced by water stress during growth. *Crop Sci.* 14: 273-78.

McCree, K. J., and Davis, S. D. 1974. Effect of water stress and temperature on leaf size and on size and number of epidermal cells in grain sorghum. *Crop Sci.* 14: 751-55.

McEwen, J. 1970. Fertilizer nitrogen and growth regulators for field beans (*Vicia faba* L.). II. The effects of large dressings of fertilizer nitrogen, single and split applications, and growth regulators. *J. Agric. Sci.* 74: 67-72.

McMichael, B. L., and Hanny, B. W. 1977. Endogenous levels of abscisic acid in water-stressed cotton leaves. *Agron. J.* 69: 979-82.

McMichael, B. L., Jordan, W. R., and Powell, R. D. 1972. An effect of water stress on ethylene production by intact cotton petioles. *Plant Physiol.* 49: 658-60.

——. 1973. Abscission processes in cotton: Induction by water deficit. *Agron. J.* 65: 202-4.

McWilliam, J. R., and Kramer, P. J. 1968. The nature of the perennial response in Mediterranean grasses. 1. Water relations and summer survival in *Phalaris. Aust. J. Agric. Res.* 19: 381-95.

Meidner, H. 1967. The effect of kinetin on stomatal opening and the rate of intake of carbon dioxide in mature primary leaves of barley. *J. Exp. Bot.* 18: 556-61.

Meidner, H., and Mansfield, T. A. 1968. *Physiology of stomata.* London: McGraw-Hill.

Meidner, H., and Sheriff, D. W. 1976. *Water and plants.* Glasgow and London: Blackie.

Meigs, P. 1953. World distribution of arid and semi-arid homoclimates. *Arid Zone Res.* 2: 203-10.

Melchior, G. L., and Morehart, A. L. 1979. Effect of water stress on the development of Verticillium wilt of yellow poplar (*Liriodendron tulipifera*). *Ann. Mtg. Potomac Div. American* 69: 536.

Milborrow, B. V., and Robinson, D. R. 1973. Factors affecting the biosynthesis of abscisic acid. *J. Exp. Bot.* 24: 537-48.

Millar, B. D., and Denmead, O. T. 1976. Water relations of wheat leaves in the field. *Agron. J.* 68: 303-7.

Millard, J. D., Jackson, R. D., Goettelman, R. C., Reginato, R. J., and Idso, S. B. 1978. Crop water stress assessment using an airborne thermal scanner. *Photo Eng. Rem. Sens.* 44: 77-85.

Miller, N. A., and Ashby, W. C. 1978. The effect of phenyl mercuric acetate on the leaf water balance of *Zea mays. Bot. Gaz.* 139: 211-14.

Milthorpe, F. L., and Moorby, J. 1974. *An introduction to crop physiology.* London: Cambridge University Press.

Mingeau, M. 1974. Comportment de colza du printemps à la secheresse. *Inf. Tech. (Paris)* 36: 1-11.

Mittelheuser, C. J., and Van Steveninck, R. F. M. 1969. Stomatal closure and inhibition of transpiration induced by (RS)-abscisic acid. *Nature* 221: 281-82.

Mizrahi, Y., Blumenfeld, A., and Richmond, A. E. 1970. Abscisic acid and transpiration in leaves in relation to osmotic root stress. *Plant Physiol.* 46: 169-71.

Mondal, R. C., and Sharma, D. R. 1979. Effect of long-term use of saline irrigation water on wheat (*Triticum aestivum*) yield and soil salinity. *Indian J. Agric. Sci.* 49: 546-50.

Monteith, J. L. 1965. Evaporation and environment. In *The state and movement of water in living organisms.* Symp. Soc. Exp. Biol. 19: 205-34. London: Cambridge University Press.

——. 1975a. *Principles of environmental physics.* London: Edward Arnold.

——. 1975b. *Vegetation and the atmosphere. Vol. 1. Principles.* New York: Academic Press.

——. 1976. *Vegetation and the atmosphere. Vol. 2. Case studies.* New York: Academic Press.

Moreshet, S. 1970. Effect of environmental factors on cuticular transpiration resistance. *Plant Physiol.* 46: 815-18.

Moreshet, S., Stanhill, G., and Fuchs, M. 1977. Effect of increasing foliage reflectance on the CO_2 uptake and transpiration resistance of a grain sorghum crop. *Agron. J.* 69: 246-50.

Morgan, J. M. 1977. Changes in diffusive conductance and water potential of wheat plants before and after anthesis. *Aust. J. Plant Physiol.* 4: 75-86.

——. 1980. Possible role for abscisic acid in reducing seed set in water-stressed wheat plants. *Nature* 285: 655-57.

Morgan, P. W., and Durham, J. I. 1975. Ethylene-induced leaf abscission is promoted by gibberellic acid. *Plant Physiol.* 55: 308-11.

Morgan, P. W., Jordan, W. R., Davenport, T. L., and Durham, J. I. 1977. Abscission responses to moisture stress, auxin transport inhibitors and ethephon. *Plant Physiol.* 59: 710-12.

Morgan, P. W., Miller, F. R., and Quinby, J. R. 1977. Manipulation of *Sorghum bicolor* growth and development with gibberellic acid. *Agron. J.* 69: 789-93.

Morilla, C. A., Boyer, J. S., and Hagemann, R. H. 1973. Nitrate reductase activity and polyribosomal content of corn (*Zea mays* L.) having low leaf water potentials. *Plant Physiol.* 51: 817-24.

Musgrave, G. W. 1955. How much of the rain enters the soil? In *Water*, ed. A. Stefferud, pp. 151-59, U.S. Dept. Agric. Yearbook.

Musick, J. T., and Grimes, D. W. 1961. Water management and consumptive use of irrigated grain sorghum in western Kansas. *Kansas Agr. Exp. Sta. Tech. Bull.* 113, 20 p.

Myers, V. I. 1970. Soil, water and plant relations. In *Remote sensing: Agriculture and forestry*. Nat. Acad. Sci. U.S. Washington, D.C.

Namken, L. N., Bartholic, J. F., and Runkles, J. R. 1969. Monitoring cotton plant stem radius as an indication of water stress. *Agron. J.* 61: 891-93.

Naylor, A. W. 1972. Water deficits and nitrogen metabolism. In *Water deficits and plant growth*, ed. T. T. Kozlowski, pp. 241-54, Vol. 3, New York: Academic Press.

Neumann, H. H., Thurtell, G. W., Stevenson, K. R., and Beadle, C. L. 1974. Leaf water content and potential in corn, sorghum, soybean and sunflower. *Can. J. Plant Sci.* 54: 185-95.

Newman, E. I. 1966. Relationship between root growth of flax (*Linum usitatissimum*) and soil water potential. *New Phytol.* 65: 273-83.

———. 1974. Root and soil water relations. In *The plant root and its environment*, ed. E. W. Carson, pp. 363-440, University Press of Virginia.

———. 1976. Interaction between osmotic and pressure-induced water flow in plant roots. *Plant Physiol.* 57: 738-39.

Nir, D. 1974. *The semi-arid world: Man on the fringe of the desert.* London: Longmans Green.

Nixon, P. R., Wiegand, C. L., Arkin, G. F., Gerberman, A. H., and Richardson, S. J. 1976. Comparisons among a grain sorghum growth model, satellite data and real world. *Bull. Amer. Meteorol. Soc.* 57: 1404.

Nour, A. E. M. and Weibel, D. E. 1978. Evaluation of root characteristics in grain sorghum. *Agron. J.* 70: 217-18.

Nowakowski, W. 1979. Influence of IAA on IAA-oxidase activity in winter wheat and maize seedlings under conditions of osmotic stress. *Acta Agrobotanica* 32: 101-7.

Obeng, L. E. 1975. Too much or too little. *Ceres* July/August: 18-20. Rome: Food and Agriculture Organization of the United Nations.

O'Neal, A. M. 1952. A key for evaluating soil permeability by means of certain field clues. *Soil. Sci. Soc. Amer. Proc.* 16: 312-15.

Ong, H. T. 1978. Roles of hormones in the responses of excised tomato cotyledons to mannitol induced water stress. *Biol. Plant.* 20: 318-23.

Oppenheimer, H. R. 1960. Adaptation to drought: xerophytism. In *Plant-water relationships in arid and semi-arid conditions.* UNESCO, Reviews of research pp. 105-38, Vol. 15. Paris.

Ordin, L. 1960. Effect of water stress on cell wall metabolism of *Avena* coleoptile tissue. *Plant Physiol.* 35: 443-50.

Osmond, C. B. 1978. Crassulacean acid metabolism: A curiosity in context. *Ann. Rev. Plant Physiol.* 29: 379-414.

Pallas, J. E., and Box, J. E. 1970. Explanation of the stomatal response of excised leaves to kinetin. *Nature* 227: 87-88.

Parker, J. 1968. Drought resistance mechanisms. In *Water deficits and plant growth*, ed. T. T. Kozlowski, pp. 195-234, Vol. 1, New York: Academic Press.

Passioura, J. B. 1972. The effect of root geometry on the yield of wheat growing on stored water. *Aust. J. Agric. Res.* 23: 745-52.

Pasternak, D., and Wilson, G. L. 1972. After-effects of night temperatures on stomatal behaviour and photosynthesis in sorghum. *New Phytol.* 71: 683-89.

——. 1974. Differing effects of water deficit on net photosynthesis of intact and excised sorghum leaves. *New Phytol.* 73: 847-50.

——. 1976. Photosynthesis and transpiration in the heads of droughted grain sorghum. *Aust. J. Exp. Agric. Anim. Husb.* 16: 272-75.

Patterson, R. F. 1965. *Wheat.* New York: Interscience Publishers.

Pavlychenko, T. K. 1937. Quantitative study of the entire root system of weeds and crop plants under field conditions. *Ecology* 18: 62-79.

Pelton, W. L. 1967. The effect of a windbreak on wind travel, evaporation and wheat yield. *Can. J. Plant Sci.* 47: 209-14.

——. 1969. Influence of low seeding rates on wheat yield in southwestern Saskatchewan. *Can. J. Plant. Sci.* 49: 607-14.

Penman, H. L. 1948. Natural evaporation from open water, bare soil and grass. *Proc. Roy. Soc.* A 193: 120-45.

——. 1949. The dependence of transpiration on weather and soil conditions. *J. Soil Sci.* 1: 74-89.

Perroux, K. M. 1979. Controlled water potential in sub-irrigated pots. *Plant and Soil* 52: 385-92.

Perween, S., and Ahmad, R. 1976. Effect of gibberellin, auxin and their inter-action on the physiology of abscission in cotton plant under normal and water stressed condition. *Pak. J. Bot.* 8: 189-98.

Philip, J. R. 1957. The physical principles of soil water movement during the irrigation cycle. *Proc. Int. Com. Irr. Drain.* 125-54.

——. 1958. The osmotic cell, solute diffusibility, and the plant water economy. *Plant Physiol.* 33: 264-71.

Philip, J. R., and de Vries, D. A. 1957. Moisture movement in porous materials under temperature gradients. *Trans. Amer. Geophys. Union* 38: 222-32.

Pierce, M., and Raschke, K. 1980. Correlation between losses of turgor and accumulation of abscisic acid in detached leaves. *Planta* 148: 174-82.

Pimentel, D., and Pimentel, M. 1977. Counting the kilocalories. *Ceres* September/October: 17-21. Rome: Food and Agriculture Organization of the United Nations.

Pitman, M. G., and Wellfare, D. 1978. Inhibition of ion transport in excised barley roots by abscisic acid: Relation to water permeability of the roots. *J. Exp. Bot.* 29: 1125-28.

Pitman, M. G., Lüttge, V., Läuchli, A., and Ball, E. 1974a. Effect of previous water stress on ion uptake and transport in barley seedlings. *Aust. J. Plant Physiol.* 1: 377-85.

——. 1974b. Action of abscisic acid on ion transport as affected by root tempera-ture and nutrient status. *J. Exp. Bot.* 25: 47-55.

Powell, D. B. B. 1974. Some aspects of water stress in late spring on apple trees. *J. Hort. Sci.* 49: 257-72.

President's Science Advisory Committee. 1974. *World food problem.* Rpt. of the President's Science Advisory Committee of the United States. Washing-ton, D. C.

Puckridge, D. W., and Donald, C. M. 1967. Competition among wheat plants sown at a wide range of densities. *Aust. J. Agric. Res.* 18: 193-211.

Pustovoitova, T. N. 1972. The effect of wilting and soil drought on the endo-genous growth regulators in mesophyllous plants. *Fiziol. Rast.* 19: 622-28.

Quarrie, S. A., and Jones, H. G. 1977. Effects of abscisic acid and water stress on development and morphology of wheat. *J. Exp. Bot.* 28: 192-203.

——. 1979. Genotype variation in leaf water potential, stomatal conductance and abscisic acid concentration in spring wheat subjected to artificial drought stress. *Ann. Bot.* 44: 323-32.

Railton, I. D., and Reid, D. M. 1973. Effects of benzyladenine on the growth of water logged tomato plants. *Planta* 111: 216-66.

Railton, I. D., Reid, D. M., Gaskin, P., and MacMillan, J. 1974. Characterization of abscisic acid in chloroplasts of *Pisum sativum* cv. Alaska by combined gas chromatography-mass spectrometry. *Planta* 117: 179-82.

Rajagopal, V., Balasubramanian, V., and Sinha, S. K. 1977. Diurnal fluctuations in relative water content, nitrate reductase and proline content in water-stressed and non-stressed wheat. *Physiol. Plant.* 40: 69-71.

Rao, N. G. P., and House, L. 1972. *Sorghum in seventies.* New Delhi, India: Oxford and IBH Publishing Co.

Raschke, K. 1975a. Stomatal action. *Ann. Rev. Plant. Physiol.* 26: 309-40.

——. 1975b. Simultaneous requirement of carbon dioxide and abscisic acid for stomatal closing in *Xanthium strumarium* L. *Planta* 125: 243-59.

——. 1976. How stomata resolve the dilemma of opposing priorities. *Phil. Trans. Roy. Soc. Lond.* B273: 551-60.

——. 1977. The stomatal turgor mechanism and its responses to CO_2 and abscisic acid: Observations and a hypothesis. In *Regulation of cell membrane activities in plants*, ed. E. Marrè and O. Cifferi, pp. 173-83. Amsterdam: Elsevier/North Holland Biomed. Press.

Raschke, K., and Zeevaart, J. A. D. 1976. Abscisic acid content, transpiration and stomatal conductance as related to leaf age in plants of *Xanthium strumarium* L. *Plant Physiol.* 58: 169-74.

Raschke, K., Furin, R. D., and Pierce, M. 1975. Stomatal closure in response to xanthoxin and abscisic acid. *Planta* 125: 149-60.

Raschke, K., Pierce, M., and Popiela, C. C. 1976. Abscisic acid content and stomatal sensitivity to CO_2 in leaves of *Xanthium strumarium* L. after pretreatment in warm and cold growth chambers. *Plant Physiol.* 57: 115-21.

Raunkaier, C. 1934. *The life form of plants and statistical plant geography.* Oxford: Clarendon.

Rawson, H. M., and Hofstra, G. 1969. Translocation and remobilization of ^{14}C assimilated at different stages by each leaf of the wheat plant. *Aust. J. Biol. Sci.* 22: 321-31.

Rawson, H. M., Begg, J. E., and Woodward, R. G. 1977. The effect of atmospheric humidity on photosynthesis, transpiration and water use efficiency of leaves of several plant species. *Planta* 134: 5-10.

Rawson, H. M., Turner, N. C., and Begg, J. E. 1978. Agronomic and physiological responses of soybean and sorghum crops to water deficits. 4. Photosynthesis, transpiration and water use efficiency. *Aust. J. Plant Physiol.* 5: 195-209.

Redmann, R. E. 1973. *Plant water relationships.* Tech. Rpt. No. 29, Matador Project (IBP), University of Saskatchewan, Saskatoon, Canada.

Reicosky, D. C., and Ritchie, J. T. 1976. Relative importance of soil resistance and plant resistance in root water absorption. *Soil Sci. Soc. Amer. J.* 40: 293-97.

Reid, D. M., and Crozier, A. 1971. Effects of water logging on the gibberellin content and growth of tomato plants. *J. Exp. Bot.* 22: 39-48.

Rennie, D. A. 1978. Our land base and what is happening to it. *Inst. for Saskatchewan Studies* 8: 2.

Richards, F. J. 1959. A flexible growth function for empirical use. *J. Exp. Bot.* 10: 290-300.

Richardson, W. N., and Stubbs, T. 1978. *Plants, agriculture and human society.* Don Mills, Ontario: W. A. Benjamin Inc.

Ripley, E. A., and Redmann, R. E. 1976. Grassland. In *Vegetation and the atmosphere. Vol. 2. Case studies,* ed. J. L. Monteith, pp. 349-98, New York: Academic Press.

Ritchie, J. T., and Arkin, G. F. 1976. Modelling leaf appearance and area development in sorghum and corn. *Bull. Amer. Meteorol. Soc.* 57: 1408.

Robertson, G. W. 1974. Wheat yields for 50 years at Swift Current, Saskatchewan in relation to weather. *Can. J. Plant Sci.* 54: 625-50.

Robins, J. S., and Domingo, C. E. 1953. Some effects of severe soil moisture deficits at specific growth stages in corn. *Agron. J.* 45: 618-21.

Rumney, G. R. 1968. *Climatology and the world's climates.* New York: MacMillan.

Russell, M. B. 1959. Water and its relation to soils and crops. *Adv. Agron.* 11: 1-131.

Salim, M. H., and Todd, G. W. 1968. Seed soaking as a pre-sowing, drought hardening treatment in wheat and barley seedlings. *Agron. J.* 60: 179-82.

Salim, M. H., Todd, G. W., and Schlehuber, A. M. 1965. Root development of wheat, oats and barley under conditions of soil moisture stress. *Agron. J.* 57: 603-7.

Saint-Clair, P. M. 1976. Germination of *Sorghum bicolor* under polyethylene glycol induced stress. *Can. J. Plant Sci.* 56: 21-24.

——. 1977. Root growth of grain-sorghum cultivars, *Sorghum bicolor* L. Moench. *Nat. Can. (Que.)* 104: 537-42.

Sanchez-Diaz, M. F., Hesketh, J. D., and Kramer, P. J. 1972. Wax filaments on sorghum leaves as seen with a scanning electron microscope. *J. Ariz. Acad. Sci.* 7: 6-7.

Schulze, E. D., Lange, O. L., Buschbom, U., Kappen, L., and Evenari, M. 1972. Stomatal responses to changes in humidity in plants growing in the desert. *Planta* 108: 259-70.

Scott, F. M. 1950. Internal suberization of tissues. *Bot. Gaz.* 111: 378-94.

Scott Russell, R. 1977. *Plant root systems: Their function and interaction with the soil.* London: McGraw-Hill.

Secretariat U.N. Conference on Desertification. 1977. *Desertification: Its causes and consequences.* Oxford, Toronto: Pergamon Press.

Sedgley, R. H., Seaton, K. A., and Stern, W. R. 1973. A field method for determining soil water availability to crops. In *Plant response to climatic factors*, ed. R. O. Slatyer, pp. 527-30, Proc. Uppsala Symp. 1970, Paris: UNESCO.

Semple, A. T. 1970. *Grassland improvement.* London: Leonard Hill.

Sharkey, T. D., and Raschke, K. 1980. Effects of phaseic acid and dihydrophaseic acid on stomata and photosynthetic apparatus. *Plant Physiol.* 65: 291-97.

Sharp, R. E., Osonubi, O., Wood, W. A., and Davies, W. J. 1979. Simple instrument for measuring leaf extension in grasses, and its application in the study of the effects of water stress on maize and sorghum. *Ann. Bot.* 44: 35-45.

Shipley, J., and Regier, C. 1970. Water response in the production of irrigated grain sorghum, High Plains of Texas, 1969. *Texas Agr. Exp. Sta. Prog. Rep.* No. 2829, 24 p.

Short, K. C., and Torrey, J. G. 1972. Cytokinins in seedling roots of pea. *Plant Physiol.* 49: 155-60.

Shukla, N. P., Singh, A. P., and Hukkeri, S. B. 1973. Effect of soil moisture stress at different stages of growth on HCN content of "MP Chari" sorghum (*Sorghum bicolor* L. Moench). *Indian J. Agric. Sci.* 43: 977-79.

Simpson, G. M. 1968. Association between grain yield per plant and photosynthetic area above the flag-leaf node in wheat. *Can. J. Plant Sci.* 48: 253-60.

Simpson, G. M., Durley, R. C., Kannangara, T., and Stout, D. G. 1979. The problem of plant breeders. In *Plant regulation and world agriculture*, ed. T. K. Scott, pp. 111-28, New York: Plenum.

Singh, T. 1978. *Drought prone areas in India*. New Delhi: People's Publishing House.

Singh, S. D., and Singh, P. 1978. Value of drip irrigation compared with conventional irrigation for vegetable production in a hot arid climate. *Agron. J.* 70: 945-47.

Singh, S. D., Gupta, J. P., and Singh, P. 1978. Water economy and saline water-use by drip irrigation. *Agron. J.* 70: 948-51.

Sivakumar, M. V. K., Seetharama, N., Singh, S., and Bidinger, F. R. 1979. Water relations, growth, and dry matter accumulation of sorghum under post-rainy season conditions. *Agron. J.* 71: 843-47.

Skidmore, E. L., Hagen, L. J., Naylor, D. G., and Teare, I. D. 1974. Winter wheat response to barrier-induced microclimate. *Agron J.* 66: 501-5.

Slabbers, P. J., Herrendorf, V. S., and Stapper, M. 1979. Evaluation of simplified water crop yield models. *Agric. Water Manage.* 2: 95-130.

Slatyer, R. O. 1957. The influence of progressive increases in soil moisture stress on transpiration, growth and internal water relationships of plants. *Aust. J. Biol. Sci.* 10: 320-36.

——. 1967. *Plant-water relationships*. London and New York: Academic Press.

Slatyer, R. O., and Taylor, S. A. 1960. Terminology in plant- and soil-water relations. *Nature* 187: 922-24.

Slatyer, R. O., and McIlroy, I. C. 1961. *Practical microclimatology*. Paris: UNESCO.

Slatyer, R. O., and Bierhuizen, J. F. 1964. The influence of several transpiration suppressants on transpiration, photosynthesis and water-use efficiency of cotton leaves. *Aust. J. Biol. Sci.* 17: 143-46.

Slavik, B. 1974. *Methods of studying plant water relations*. Berlin, New York: Springer Verlag.

Smith, B. F., and Aldrich, D. T. A. 1967. Spring bean variety trials 1954/65. *J. Natn. Inst. Agric. Bot.* 11: 114-32.

Sofield, I., Evans, L. T., Cooke, M. G., and Wardlaw, I. F. 1977. Factors influencing rate and duration of grain filling in wheat. *Aust. J. Plant Physiol.* 4: 785-97.

Spence, L. E. 1937. Root studies of important range plants of the Boise river watershed. *J. Forestry* 35: 747-54.

Sprent, J. I. 1972a. The effect of water stress on nitrogen-fixing root nodules. II. Effects on the fine structure of detached soybean nodules. *New Phytol.* 71: 443-50.

——. 1972b. The effects of water stress on nitrogen-fixing root nodules. IV. Effects on whole plants of *Vicia faba* and *Glycine max. New Phytol.* 71: 603-11.

Sprent, J. I., and Bradford, A. M. 1977. Nitrogen fixation in field beans (*Vicia faba*) as affected by population density, shading and its relationship with soil moisture. *J. Agric. Sci.* 86: 303-10.

Sprent, J. I., Bradford, A. M., and Norton, C. 1977. Seasonal growth patterns in field beans (*Vicia faba*) as affected by population density, shading and its relationship with soil moisture. *J. Agric. Sci.* 88: 293-301.

Sprent, J. I., Bradford, A. M., and Gallacher, A. E. 1978. Factors affecting nodulation and nitrogen fixation by *Vicia faba. Ann. Appl. Biol.* 85: 473-76.

Srivastava, A. K., and Singh, G. 1979. Pre-sowing hardening treatment of Lentil (*Lens culinaris*) seeds for inducing resistance to moisture stress. *Indian J. Exp. Biol.* 17: 1280-84.

Stanhill, G. 1976. Cotton. In *Vegetation and the atmosphere Vol. 2. Case studies*, ed. J. L. Monteith, pp. 121-50, London: Academic Press.

Stanhill, G., Moreshet, S., and Fuchs, M. 1976. Effect of increasing foliage and soil reflectivity on the yield and water use efficiency of grain sorghum. *Agron. J.* 68: 329-32.

Staple, W. J. 1960. Significance of fallow as a management technique in Continental and winter-rainfall climates. In *Plant water relationships in arid and semi-arid conditions*, Vol. 15, Paris: UNESCO.

Staple, W. J., and Lehane, J. J. 1954. Weather conditions influencing wheat yields in tanks and field plots. *Can. J. Agric. Res.* 34: 552-64.

St. John, J. B., and Christiansen, M. N. 1977. Modification of plant response to temperature stress with a substituted pyradazinone BASF-13-338. *Proc. Northeast. Weed Sci. Soc.* 31: 144.

Stone, L. R., Teare, I. D., Nickell, C. D., and Mayaki, W. C. 1976. Soybean root development and soil water depletion. *Agron. J.* 68: 677-80.

Stout, D. G., and Simpson, G. M. 1978. Drought resistance of *Sorghum bicolor.*

1. Drought avoidance mechanisms related to leaf water status. *Can. J. Plant. Sci.* 58: 213-24.

Stout, D. G., Kannangara, T., and Simpson, G. M. 1978. Drought resistance of *Sorghum bicolor.* 2. Water stress effects on growth. *Can. J. Plant Sci.* 58: 225-33.

Stout, D. G., Simpson, G. M., and Flotre, D. M. 1980. Drought resistance of *Sorghum bicolor* L. Moench. 3. Seed germination under osmotic stress. *Can. J. Plant Sci.* 60: 13-24.

Sturrock, J. W. 1975. Wind effects and their amelioration in crop production. In *Physiological aspects of dryland farming,* ed. U. S. Gupta, pp. 285-313, New Delhi: Oxford and IBH Publishing.

Suh, H. W., Dayton, A. D., Casady, A. J., and Liang, G. H. 1976. Diallel cross analysis of stomatal density and leaf blade area in grain sorghum (*Sorghum bicolor*). *Can. J. Genet. Cytol.* 18: 679-86.

Sullivan, C. Y. 1972. Mechanisms of heat and drought resistance in grain sorghum and methods of measurement. In *Sorghum in seventies,* ed. N. G. P. Rao and L. R. House, pp. 247-64, New Delhi, India: Oxford and IBH Publishing.

Sullivan, C. Y., and Eastin, J. D. 1974. Plant physiological responses to water stress. *Agric. Meteorol.* 14: 113-27.

Sullivan, C. Y., and Ross, W. M. 1979. Selecting for drought and heat resistance in grain sorghum. In *Stress physiology in crop plants,* ed. H. Mussell and R. C. Staples, pp. 262-81, New York: Wiley.

Sumayo, C. R. Kanemasu, E. T., and Hodges, T. 1977. Soil moisture effects on transpiration and net carbon dioxide exchange of sorghum. *Agric. Meteorol.* 18: 401-8.

Sung, F. J. M., and Krieg, D. R. 1979. Relative sensitivity of photosynthetic assimilation and translocation of carbon-14 to water stress. *Plant Physiol.* 64: 852-56.

Szeicz, G., van Bavel, C. H. M., and Takami, S. 1973. Stomatal factor in the water use and dry matter production by sorghum. *Agric. Meteorol.* 12: 361-89.

Tal, M., and Imber, D. 1970. Abnormal stomatal behaviour and hormonal imbalance in *Flacca,* a wilty mutant of tomato. II. Auxin and abscisic acid-like activity. *Plant Physiol.* 46: 373-76.

——. 1971a. Abnormal stomatal behaviour and hormonal imbalance in *Flacca,* a wilty mutant of tomato. III. Hormonal effects on the water status in the plant. *Plant Physiol.* 47: 849-50.

——. 1971b. The effect of a prolonged, 2,4-dichlorophenoxyacetic acid treatment on transpiration and stomatal distribution in tomato leaves. *Planta* 97: 179-82.

Tanner, C. B. 1960. Energy balance approach to evapotranspiration from crops. *Soil Sci. Soc. Amer. Proc.* 24: 1-9.

——. 1968. Evaporation of water from plants and soil. In *Water deficits and plant growth*, ed. T. T. Kozlowski, pp. 73-106, New York, San Francisco: Academic Press.

Tarabrin, V. P., and Teteneva, T. R. 1979. Presowing treatment of seeds and its effect on the resistance of woody plants against drought. *Sov. J. Ecol.* 10: 204-11.

Tateno, K., and Ojima, M. 1976. Effects of temperature and soil water content during grain filling period on the yields of grain sorghum. *Proc. Crop Sci. Soc. Japan* 45: 63-68.

Taylor, H. M., and Klepper, B. 1971. Water uptake by cotton roots during an irrigation cycle. *Aust. J. Biol. Sci.* 24: 853-59.

——. 1973. Rooting density and water extraction patterns for corn (*Zea mays* L.). *Agron. J.* 65: 965-68.

——. 1974. Water relations of cotton. 1. Root growth and water use as related to top growth and soil water content. *Agron. J.* 66: 584-88.

——. 1978. The role of rooting characteristics in the supply of water to plants. *Adv. Agron.* 30: 99-128.

Teare, I. D., Kanemasu, E. T., Powers, W. L., and Jacobs, H. S. 1973. Water use efficiency and its relation to crop canopy area, stomatal regulation and root distribution. *Agron. J.* 65: 207-11.

Teare, I. D., Schimmelpfennig, H., and Waldren, R. P. 1973. Rainout shelter and drainage lysimeters to quantitatively measure drought stress. *Agron. J.* 65: 544-49.

Thornthwaite, C. W., 1948. An approach toward a rational classification of climate. *Geogr. Rev.* 38: 55-94.

Thornthwaite, C. W., and Holzman, B. 1939. The determination of evaporation from land and water surfaces. *Mon. Weath. Rev.* 67: 4-11.

Ting, I. P., 1976. Crassulacean acid metabolism in natural ecosystems in relation to annual CO_2 uptake patterns and water utilization. In *CO_2 metabolism and plant productivity*, ed. R. H. Burris and C. C. Black, pp. 251-68, Baltimore: University Park Press.

Tinklin, R., and Weatherley, P. E. 1966. On the relationship between transpiration rate and leaf water potential. *New Phytol.* 65: 509–17.

Todd, G. W., and Webster, D. L. 1965. Effects of repeated drought periods on photosynthesis and survival of cereal seedlings. *Agron. J.* 57: 399–404.

Townley-Smith, T. F., and Hurd, E. A. 1979. Testing and selecting for drought resistance in wheat. In *Stress physiology in crop plants*, ed. H. Mussell and R. C. Staples, pp. 447–64, New York: Wiley-Interscience.

Troughton, J. H., Bjoerkman, O., and Perry, J. A. 1974. Growth of sorghum at high temperatures. In *Carnegie Institute of Washington Year Book 1973*, ed. S. A. McDough, pp. 835–38, Washington, D.C.: Carnegie Institute of Washington.

Tucker, G. J. 1979. Rev. and photo IK linear combinations for monitoring vegetation. *Rem. Sens. Env.* 8: 127–50.

Turner, N. C. 1974. Stomatal behaviour and water status of maize, sorghum and tobacco under field conditions. 2. Low soil water potential. *Plant Physiol.* 53: 360–65.

——. 1979. Drought resistance and adaptation to water deficits in crop plants. In *Stress physiology in crop plants*, ed. W. C. Mussell and R. C. Staples, pp. 343–72, New York: Wiley.

Turner, N. C., and Begg, J. E. 1973. Stomatal behaviour and water status of maize sorghum and tobacco under field conditions. 1. High soil water potential. *Plant Physiol.* 51: 31–36.

Turner, N. C., Begg, J. C., and Tonnet, M. L. 1978. Osmotic adjustment of sorghum and sunflower crops in response to water deficits and its influence on the water potential at which stomata close. *Aust. J. Plant Physiol.* 5: 597–608.

Tyree, M. T. 1976. Negative turgor pressure in plant cells: Fact or fallacy? *Can. J. Bot.* 54: 2738–46.

Uhvits, R. 1946. Effect of osmotic pressure on water absorption and germination of alfalfa seeds. *Am. J. Bot.* 33: 278–85.

Unger, P. W. 1978. Straw mulch effects on soil temperature and *Sorghum bicolor* germination and growth. *Agron. J.* 70: 858–64.

United States Department of Agriculture. 1974. *Agricultural statistics 1974*. Government Printing Office. Washington, D.C.

United States Department of Agriculture. 1975. *National food situation 1975*. U.S. Dept. Agr. Econ. Res. Serv. NFS-151.

Vaadia, Y., Raney, F. C., and Hagan, R. M. 1961. Plant water deficits and physiological processes. *Ann. Rev. Plant Physiol.* 12: 265-92.

Van Arkel, H. 1978. Leaf area determinations in sorghum and maize by the length width method. *Neth. J. Agric. Sci.* 26: 170-80.

Van Bavel, C. H. M., and Ahmed, J. 1976. Dynamic simulation of water depletion in the root zone. *Ecol. Model.* 2: 189-212.

Van den Honert, T. H. 1948. Water transport in plants as a catenary process. *Disc. Farad. Soc.* 3: 146-53.

Van der Boon, J. 1973. Influence of potassium calcium ratio and drought on physiological disorders in tomato. *Neth. J. Agric. Sci.* 21: 56-67.

Vanderlip, R. L., and Arkin, G. F. 1977. Simulating accumulation and distribution of dry matter in grain sorghum. *Agron. J.* 69: 917-23.

Van Kirk, C. A., and Raschke, K. 1978. Release of malate from epidermal strips during stomatal closure. *Plant Physiol.* 61: 474-75.

Van Steveninck, R. F. M. 1972. Abscisic acid stimulation of ion transport and alteration in K^+/Na^+ selectivity. *Z. Pflanzenphysiol.* 67: 282-86.

Vavilov, N. I. 1949-50. The origin, variation, immunity and breeding of cultivated plants. *Chronica Botanica* 13: 1-366.

Verma, S. B., and Rosenberg, N. J. 1977. Brown-Rosenberg resistance model of crop evapotranspiration modified tests in an irrigated sorghum field. *Agron. J.* 69: 332-35.

Von Bertalanffy, L. 1968. *General system theory.* New York: George Braziller.

Waddell, E. 1977. The return of traditional agriculture: The only means of solving the world food problem. *The Ecologist* (UK) 7: 144-47.

Waldren, R. P., Teare, I. D., and Ehler, S. W. 1974. Changes in free proline concentration in sorghum and soybean plants under field conditions. *Crop Sci.* 14: 447-50.

Wallace, D. H., Ozbun, J. L., and Munger, H. M. 1972. Physiological genetics of crop yield. *Adv. Agron.* 24: 97-146.

Walter, H. 1973. *Vegetation of the earth.* New York: Springer-Verlag.

Walter, H., and Stadlemann, E. 1974. A new approach to the water relations of desert plants. In *Desert biology*, Vol. 2, ed. G. W. Brown, pp. 214-310, New York: Academic Press.

Walton, D. C. 1980. Biochemistry and physiology of abscisic acid. *Ann. Rev. Plant Physiol.* 31: 453-89.

Walton, D. C., Harrison, M. A., and Coté, P. 1976. The effects of water stress on abscisic acid levels and metabolism in roots of *Phaseolus vulgaris* L. and other plants. *Planta* 131: 141-44.

Walton, D. C., Galson, E., and Harrison, M. A. 1977. The relationship between stomatal resistance and abscisic acid levels in leaves of water-stressed bean plants. *Planta* 133: 145-48.

Wardlaw, I. F. 1969. The effect of water stress on translocation in relation to photosynthesis and growth. 2. Effect during leaf development in *Lolium perenne* L. *Aust. J. Biol. Sci.* 22: 1-16.

——. 1976. Assimilate partitioning: Cause and effect. In *Transport and transfer processes in plants*, ed. I. F. Wardlaw and J. B. Passioura, pp. 381-91, New York: Academic Press.

Wareing, P. F., Horgan, R., Henson, I. E., and Davis, W. 1977. Cytokinin relations in the whole plant. In *Plant growth regulation*, ed. P. E. Pilet, pp. 147-53. Berlin, Heidelberg, New York: Springer-Verlag.

Warren Wilson, J. 1967a. The components of leaf water potential. I. Osmotic and metric potentials. *Aust. J. Biol. Sci.* 20: 329-47.

——. 1967b. The components of leaf water potential. II. Pressure potential and water potential. *Aust. J. Biol. Sci.* 20: 349-57.

——. 1967c. The components of leaf water potential. III. Effects of tissue characteristics and relative water content on water potential. *Aust. J. Biol. Sci.* 20: 359-67.

Weatherley, P. E. 1963. The pathway of water movement across the root cortex and leaf mesophyll of transpiring plants. In *The water relations of plants*, ed. A. J. Rutter and F. H. Whitehead, pp. 85-100, London: Blackwell.

——. 1965a. Some investigations on water deficit and transpiration under controlled conditions. In *Water stress in plants*, Proc. Sympos. Prague, ed. B. Slavik, pp. 63-71. The Hague: Dr. W. Junk.

——. 1965b. Water in the leaf. In *The state and movement of water in living organisms*. Symp. Soc. Exp. Biol. 19: 157-84, London: Cambridge University Press.

——. 1976. Water movement through plants. *Phil. Trans. Roy. Soc. Lond.* B273: 435-44.

Weaver, J. E. 1926. *Root development of field crops*. New York: McGraw-Hill.

Wellburn, A. R., Ogunkanmi, A. B., Fenton, R., and Mansfield, T. A. 1974.

All-*trans*-farnesol: A naturally occurring anti transpirant? *Planta* 120: 255-63.

Wetmore, R. H., and Jacobs, W. O. 1953. Studies on abscission. The inhibition effect of auxin. *Am. J. Bot.* 40: 272-76.

Weyers, J. D. B., and Hillman, J. R. 1979. Uptake and distribution of abscisic acid in *Commelina* leaf epidermis. *Planta* 144: 167-72.

——. 1980. Effects of abscisic acid on $^{86}Rb^{+}$ fluxes in *Commelina-communis* L. leaf epidermis. *J. Exp. Bot.* 31: 711-20.

Wiese, A. F., and Army, T. J. 1958. Effect of tillage and chemical weed control practices in soil moisture storage and losses. *Agron. J.* 50: 465-68.

Wiggans, S. C., and Gardner, F. P. 1959. Effectiveness of various solutions for simulating drought conditions as measured by germination and seedling growth. *Agron. J.* 51: 315-18.

Willat, S. T., and Taylor, H. M. 1978. Water uptake by soya-bean roots as affected by their depth and by soil water content. *J. Agric. Sci.* 90: 205-13.

Williams, J. 1976. Dependence of root water potential on root radius and density. *J. Exp. Bot.* 27: 121-24.

Williams, E. A., and Morgan, P. W. 1979. Floral initiation in sorghum hastened by giberellic acid and far-red light. *Planta* 145: 269-72.

Willis, W. O., Haas, H. J., and Carlson, C. W. 1969. Snowpack runoff as affected by stubble height. *Soil Sci.* 107: 256-59.

Wilson, A. J., Robards, A. W., and Goss, M. J. 1978. Effects of mechanical impedence on barley roots—changes in ultrastructure. *Letcombe Lab. Ann. Rep. 1977*, pp. 25-7.

Wittwer, S. H. 1978. The next generation of agricultural research. *Science* 199: 375.

Woodruff, D. R. 1969. Studies on presowing drought hardening of wheat. *Aust. J. Agric. Res.* 20: 13-24.

Wright, S. T. C. 1969. An increase in the "inhibitor β" content of detached wheat leaves following a period of wilting. *Planta* 86: 10-20.

——. 1977. The relationship between leaf water potential (ψ_{leaf}) and the levels of abscisic acid and ethylene in excised wheat leaves. *Planta* 134: 183-89.

——. 1979. The effect of 6-benzyladenine and leaf ageing treatment on the levels of stress induced ethylene emanating from wilted wheat leaves. *Planta* 144: 179-88.

———. 1980. The effect of plant growth regulator treatments on the levels of ethylene emanating from excised turgid and wilted wheat *Triticum aestivum* leaves. *Planta* 148: 381–88.

Wright, S. T. C., and Hiron, R. W. P. 1969. (+)-abscisic acid, the growth inhibitor in detached wheat leaves following a period of wilting. *Nature* 224: 719–20.

Yamamoto, W. S. 1965. Homeostasis, continuity, and feedback. In *Physiological controls and regulations*, ed. W. S. Yamamoto and J. R. Brobeck, Philadelphia: W. B. Saunders.

Youngs, E. G. 1958. Redistribution of moisture in porous materials after infiltration. *Soil Sci.* 86: 117–25.

———. 1965. Water movement in soils. In *The state and movement of water in living organisms*. Symp. Soc. Exp. Biol. 19: 89–112, London: Cambridge University Press.

Zabadal, T. J. 1974. A water potential threshold for the increase of abscisic acid in leaves. *Plant Physiol.* 53: 125–27.

Zahner, R. 1968. Water deficits and growth of trees. In *Water deficits and plant growth*, ed. T. T. Kozlowski, pp. 191–254, Vol. 2, New York: Academic Press.

Zartman, R. E., and Woyewodzic, R. T. 1979. Root distribution patterns of two hybrid grain sorghums under field conditions. *Agron. J.* 71: 325–28.

Zeevaart, J. A. D. 1974. Levels of (+)-abscisic acid and xanthoxin in spinach under different environmental conditions. *Plant Physiol.* 53: 644–48.

INDEX